MEDIA AND COMMUNICATIONS - TECHNOLOGIES, POLICIES AND CHALLENGES

EMERGENCY COMMUNICATIONS

POLICY, TECHNOLOGY AND FUNDING CONSIDERATIONS

Media and Communications - Technologies, Policies and Challenges

Additional books in this series can be found on Nova's website under the Series tab.

Additional E-books in this series can be found on Nova's website under the E-books tab.

Safety and Risk in Society

Additional books in this series can be found on Nova's website under the Series tab.

Additional E-books in this series can be found on Nova's website under the E-books tab.

MEDIA AND COMMUNICATIONS - TECHNOLOGIES, POLICIES AND CHALLENGES

EMERGENCY COMMUNICATIONS

POLICY, TECHNOLOGY AND FUNDING CONSIDERATIONS

DANIEL H. MARTINEZ
AND
ALLEN E. YOUNG
EDITORS

Nova Science Publishers, Inc.
New York

Library of Congress Cataloging-in-Publication Data

ISBN 978-1-62081-140-5

Published by Nova Science Publishers, Inc. † New York

CONTENTS

PREFACE

The United States has yet to find a solution that assures seamless communications among first responders and emergency personnel at the scene of a major disaster. Since September 11, 2001, when communications failures contributed to the tragedies of the day, Congress has passed several laws intended to create a nationwide emergency communications capability. The 111th Congress considered pivotal issues, such as radio frequency spectrum license allocation and funding programs for a Public Safety Broadband Network (PSBN), without finding a solution that satisfied the expectations of both public safety and commercial network operators. This book explores the areas where changes in existing policies and practices may facilitate achievement of the important goals for emergency communications that Congress and others have identified. Why these goals have become important, and recent planning efforts to achieve these goals is discussed.

Chapter 1- The United States has yet to find a solution that assures seamless communications among first responders and emergency personnel at the scene of a major disaster. Since September 11, 2001, when communications failures contributed to the tragedies of the day, Congress has passed several laws intended to create a nationwide emergency communications capability. The 111th Congress considered pivotal issues, such as radio frequency spectrum license allocation and funding programs for a Public Safety Broadband Network (PSBN), without finding a solution that satisfied the expectations of both public safety and commercial network operators. Congressional initiatives to advance public policies for Next Generation 911 services (NG9-1-1) also remained incomplete. The 112th Congress is under renewed pressure to come to a decision about the assignment of a block of radio frequency spectrum licenses referred to as the D Block, and to provide a plan for federal support of broadband networks for emergency communications. The cost of constructing new networks (wireless and wireline) is estimated by experts to be in the tens of billions of dollars over the long term, with similarly large sums needed for maintenance and operation. Identifying money for federal support in the current climate of budget constraints provides a challenge to policy makers. The greater challenge, however, may be to assure that funds are spent effectively toward the national goals that Congress sets.

Chapter 2- Effective emergency response is dependent on wireless communications. To minimize communications failures during and after a crisis requires ongoing improvements in emergency communications capacity and capability. The availability of radio frequency spectrum is considered essential to developing a modern, interoperable communications network for public safety. Also critical are (1) building the network to use this spectrum and (2) developing and deploying the radios to the new standards required for mobile broadband. Beyond recognition of these common needs and goals, opinions diverge on such issues as how much spectrum should be made available for public safety broadband communications, how communications networks should be configured, who should own them, who should build them, who should operate them, who should be allowed to use them, and how they might be paid for.

Chapter 3- In March 2010, the FCC released its National Broadband Plan (NBP), which made significant recommendations for improving access to broadband communications across America and for enhancing the role of broadband in public safety and emergency response.

In particular, the NBP proposed a comprehensive strategy for creating a nationwide interoperable public safety broadband wireless network ("public safety broadband network") for first responders and other public safety personnel. This strategy includes:

Creating an administrative system that ensures access to sufficient capacity on a day-to-day and emergency basis;

Ensuring there is a mechanism in place to promote interoperability and operability of the network; and

Establishing a funding mechanism to ensure the network is deployed throughout the United States and has necessary coverage, resiliency and redundancy.

Chapter 4- . I am Jack Parow, recently retired chief of the Chelmsford (MA) Fire Rescue Department and president and chairman of the board of the

International Association of Fire Chiefs (IAFC) on whose behalf I appear. My organization represents the leadership of over 1.2 mllion firefighters and emergency responders. IAFC members are the world's leading experts in firefighting, emergency medical services, terrorism response, hazardous materials spills, natural disasters, search and rescue, and public safety policy. As far back as 1873, the IAFC has provided a forum for members to exchange ideas and find the latest products and services available to first responders.

I testify today in support of H.R. 607 (Broadband for First Responders Act of 2011). Currently, this legislation has eleven cosponsors and is bipartisan. Obviously, this is not part of a political agenda. It is legislation to benefit public safety agencies and the citizens whom they serve and protect. We are grateful for the efforts of this committee to keep this issue foremost in congressional consideration. Public safety is an integral part of this nation's homeland security. Local fire, emergency medical services (EMS) and law enforcement constitute the first response to incidents both small and large--natural, accidental and illicit. This legislation is needed to enable fire, EMS and law enforcement to reach our highest goal – construction of a nationwide public safety wireless interoperable broadband network.

In: Emergency Communications ISBN: 978-1- 62081-140-5
Editors: D. H. Martinez and A. E. Young © 2012 Nova Science Publishers, Inc.

Chapter 1

FUNDING EMERGENCY COMMUNICATIONS: TECHNOLOGY AND POLICY CONSIDERATIONS*

Linda K. Moore

SUMMARY

The United States has yet to find a solution that assures seamless communications among first responders and emergency personnel at the scene of a major disaster. Since September 11, 2001, when communications failures contributed to the tragedies of the day, Congress has passed several laws intended to create a nationwide emergency communications capability. The 111th Congress considered pivotal issues, such as radio frequency spectrum license allocation and funding programs for a Public Safety Broadband Network (PSBN), without finding a solution that satisfied the expectations of both public safety and commercial network operators. Congressional initiatives to advance public policies for Next Generation 911 services (NG9-1-1) also remained incomplete. The 112th Congress is under renewed pressure to come to a decision about the assignment of a block of radio frequency spectrum licenses referred to as the D Block, and to provide a plan for

* This is an edited, reformatted and augmented version of a Congressional Research Service publication, CRS Report for Congress R41842, from www.crs.gov, prepared for Members and Committees of Congress, dated January 5, 2012.

federal support of broadband networks for emergency communications. The cost of constructing new networks (wireless and wireline) is estimated by experts to be in the tens of billions of dollars over the long term, with similarly large sums needed for maintenance and operation. Identifying money for federal support in the current climate of budget constraints provides a challenge to policy makers. The greater challenge, however, may be to assure that funds are spent effectively toward the national goals that Congress sets.

After years of debate, a majority in the public safety community has agreed to implement common technologies using Internet Protocol (IP)-enabled networks and the wireless technology known as Long Term Evolution (LTE) to build a nationwide PSBN. IP-enabled networks are also considered essential to the introduction of NG9-1-1. The adoption of the Internet Protocol for emergency communications represents a significant advance in the technologies available for response and recovery operations. IP-enabled technologies are faster and smarter, capable of analyzing and directing communications as they move through networks. Achieving the transition to a leading-edge, broadband network powered by the next generation of IP technologies requires significant changes in operations and long-standing agency traditions, major investments in infrastructure and radios, and the development of enabling technologies.

The need appears increasingly urgent for timely decisions by policy makers on new infrastructure for emergency communications and spectrum allocation for public safety radios. Commercial deployment of wireless networks using LTE standards that might also support public safety use are out-pacing the planning efforts of public safety and government officials.

The House of Representatives, on December 13, 2011, approved the Middle Class Tax Relief and Job Creation Act of 2011 (H.R. 3630, Representative Camp). H.R. 3630, Title IV contains provisions, approved in markup on December 1, 2011, by the Subcommittee on Communications and Technology, House Committee on Energy and Commerce. H.R. 3630 is under consideration by a conference committee for which a major focus of discussion is an extension of payroll tax cuts and how to fund them. On the Senate side, the legislative response to H.R. 3630 will likely include provisions from the American Jobs Act of 2011 (S. 1549 and others) and from the Public Safety and Wireless Innovation Act (S. 911, as amended, Senator Rockefeller). S. 911 received bipartisan approval by the Committee on Commerce, Science, and Transportation.

Other legislation that has been introduced in the 112th Congress to address some of these issues includes the Broadband for Public Safety Act of 2011 (S. 1040, Lieberman), and the Broadband for First Responders Act (H.R. 607, King).

GOALS AND CHALLENGES

For nearly a decade, Congress and successive administrations have worked to improve emergency communications across the nation.[1] Although the issue of interoperable communications for first responders[2] has drawn the most attention, there have also been several initiatives, including legislation, to upgrade the functionality of 911 call centers. Activities marking the 10th anniversary of September 11, 2001, increased pressure on Congress to provide a comprehensive solution for better emergency communications to meet both local public safety needs and national goals. The 112th Congress has examined—and is expected to continue to examine—assuring nationwide availability of state-of-the art communications capability for emergency response and recovery.[3]

If there is common agreement among most public safety and government officials, industry leaders, and policy makers for the need to act, a plan for action has not been agreed upon. The White House has supported allocating additional radio frequency spectrum and funding in support of a new, wireless public safety communications network.[4] This position has received widespread support among the public safety community and some commercial stakeholders, such as the Public Safety Alliance.[5] Proposed federal funding for this solution, however, has raised concerns about the cost to the government. The Federal Communications Commission (FCC) has proposed a public-private partnership that would use radio frequency spectrum to attract a commercial partner willing to share network infrastructure with public safety entities, reducing costs but also limiting the amount of spectrum capacity under the direct control of public safety agencies.[6] The FCC's plans have received support from the commercial sector and stakeholders and policy makers who believe that it will increase competition and lower costs to all users of wirelessnetworks.[7] The number and variety of conflicting interests have made it difficult for Congress to reach agreement. Of the bills introduced in the 112th Congress, as of the date of this report, S. 911, as amended, and the amended Discussion Draft of the Jumpstarting Opportunity with Broadband Spectrum (JOBS) Act of 2011,[8] which has been incorporated in H.R. 3630, have been reported favorably by committee.

This report identifies areas where changes in existing policies and practices may facilitate achievement of the important goals for emergency communications that Congress and others have identified. Why these goals have become important, and recent planning efforts to achieve these goals, is discussed first. Next, possible barriers to achieving these goals are identified and described. The conclusion revisits key options presently under consideration by Congress.

Background: Planning to Meet Goals

Prior to September 11, 2001, planning for emergency communications rested largely with states and communities, with an emphasis on effective response by different types of service (police, fire, ambulance) at the local level. As radio technologies evolved, there was a growing awareness among first responders—long before 9/11—of the need for better coordination and communications interoperability.[9] Their concerns crystallized after a commercial aircraft crashed into a bridge crossing the Potomac River between Washington, DC, and Arlington County, Virginia, in January 1982.[10] As a result, new planning efforts for multi-jurisdictional and multidisciplinary responses were put in place across the United States, with federal agencies often providing assistance through technical expertise and funding.

In 1996, a report was published by the Public Safety Wireless Advisory Committee (PSWAC), which the Federal Communications Commission and the National Telecommunications and Information Administration (NTIA) had created to provide Congress with recommendations on meeting the communications needs of first responders. The committee's *Final Report*[11] was considered a landmark in planning and goal setting for public safety communications. Post-9/11, the concerns raised by PSWAC received new attention. In 2003, a National Task Force on Interoperability, with the support of the Department of Justice, revisited the PSWAC report findings, reiterating needs such as investment in infrastructure, more funding, better planning, more coordination and cooperation among agencies, and allocation of additional radio frequency spectrum.[12] In 2007, Congress created the Office of Emergency Communications (OEC) within the Department of Homeland Security (DHS) and tasked it with the preparation of a National Emergency Communications Plan (NECP), which was published in 2008.[13] Separately, Congress, also in 2008, required the National E9-1-1 Implementation Coordination Office (ICO)[14] to prepare a national plan for migration to Next Generation 911 (NG9-1-1) and to identify possible actions for Congress to take in advancing goals identified through the planning process.[15] The FCC's 2010 *National Broadband Plan*[16] included goals for advancing "robust and secure public safety communications networks."[17] All of these planning efforts included a broad base of stakeholders in the process. The plans, separately compiled and administered, have provided much of the basis for current federal policy and technology strategies. The common thread is the stakeholders who have participated in the various planning committees; for example, key public safety associations are consistently represented on these committees.

The FCC's broadband plan advocated deploying IP-enabled technologies for NG9-1-1 and for wireless broadband networks for public safety radios. Planning efforts for both of these emergency communications have since been initiated by the FCC.[18]

Challenges and Barriers to Change

Although many of the plans developed by federal agencies over the last decade have encouraged statewide and regional planning, the evolutionary path for providing emergency communications has favored local planning over federal, regional, and even state authority. In general, the larger cities and more prosperous counties have taken the lead in deploying new communications systems. Cities and towns with populations of less than 100,000 and rural counties apparently may struggle with implementing modern communications systems. Lack of funds to build infrastructure may not be the only difficulty; often these communities do not have the needed personnel, such as grants administrators or information technology managers, and therefore rely on sharing resources with neighboring counties.

The Department of Homeland Security has advocated emergency communications planning from the bottom up, encouraging stakeholders to find their own solutions within frameworks established within DHS, evolving along a development continuum provided by the agency.[19] Although localism offers many benefits for communities, the relative autonomy in choosing technology has led to incompatible equipment and small-scale, inefficient markets for equipment and infrastructure. Emergency communications technology choices have tended to be proprietary, tailored to specific needs of micro-markets, and closed to commercial, mass-market solutions.[20] As a consequence, there is a notable gap between the level of technology available to the typical first responder or 911 call center operator and that used every day by the average American consumer.

Today, emergency communications systems are typically local in reach and limited in applications. According to most experts, the emergency communications grid of the not-toodistant future is likely to be local, national, and international in reach, with flexible and accessible applications that can be instantaneously tailored to meet the needs of any emergency situation. The emergency communications technologies of the future are expected to be developed for common IP-enabled platforms that can operate on any IP-enabled network. IP-enabled wireless devices using technologies such as Long Term Evolution (LTE) will deliver advanced services anywhere, any time.[21]

Implementation of the next generation of emergency communications technology may alter the patterns of the past 30 years of technology development and adoption for public safety radios and systems. The transition might be facilitated by changes in the management of public safety resources, addressing questions such as the following:

- Who will be responsible for governance?
- Where will decision making and planning reside?
- How will reliable cost estimates be established?
- What sources of revenue are available?
- How will radio frequency spectrum assets be managed?
- How will technology be developed and applied?

Policy Considerations for the Future

Current management of public safety resources—along with some comments about how these may be changed in the future—is discussed below in six sections: governance, decision making, cost estimates, sources of revenue, spectrum management, and technology. Current policies in these areas may be inadequate to achieve the desired levels of emergency capabilities for response and recovery in all regions of the country. Congress may therefore choose to consider these and other areas where policy recommendations or legislative actions may expedite improvements to emergency communications.

Governance and Grants

Governance is a broad term that can include a number of factors. At the federal level, governance is closely associated with evaluation of existing programs and the management of federal expenditures, including grants. A number of federal agencies have roles in guiding and monitoring some decisions of states and localities through grant administration. Currently, over 40 active programs, administered by nine different departments and multiple agencies within those departments, are providing grants for funding emergency communications, and guidance and governance for those grants.[22] Within DHS, the Office of Emergency Communications, the SAFECOM Program, and the Federal Emergency Management Administration (FEMA) are among the agencies that formulate policies, plan exercises, provide guidelines, and establish requirements.[23] Two major grants programs within the Department of Justice provide funds for emergency communications; these are the Community Oriented Policing Services (COPS) program and the Edward Byrne Memorial Justice Assistance Grant (JAG) program.[24] The NTIA administers grants made through the Broadband Technologies Opportunity Program (BTOP), some of which benefit public safety communications and 911 services.[25] The NTIA shared responsibility with the Department of Transportation, National Highway Traffic Safety Administration, for a 911 grant program[26] established by the ENHANCE 911 Act of 2004.[27] The only federal grant program exclusively for improving 911, its sole appropriations was less than $50 million provided through the Digital Transition and Public Safety Fund, created by the Deficit Reduction Act of 2005.[28]

It is estimated that as much as $13 billion in federal funding, possibly more, was spent on emergency communications from FY2001 through FY2010. This total does not take full account of earmarks and other forms of assistance not reported as grants. Little information about how this money was spent is readily available.[29]

Grant programs provide a mechanism for governance by affecting performance and behavior; financial audits and other reviews of expenditures provide oversight. For investments in public safety communications systems and equipment, oversight seems to be mostly within the communities whose first responders use the systems that have been purchased. Federal audits usually are not performed at a level of detail that identifies the specific uses of federal grants by the ultimate recipients of the funds.

Because new technologies that might benefit public safety communications are being developed and deployed almost exclusively within the commercial sector, one of the considerations for improving governance of emergency communications deployments might be how to manage partnerships between the commercial sector and state and local agencies. In a publicly traded corporation, governance is provided by the board of directors and senior management, who must satisfy the expectations of shareholders and financial markets as well as assure compliance with rules and regulations. A public-private partnership might therefore meld public and private methods of governance. This in turn might lead to a closer examination of the federal role in providing governance and funding. For example, to capture the full advantages from innovation and cost-saving economies of scale, some say transfers of new commercial technologies to public safety agencies should occur on an on-going basis, not as episodic one-time transactions. Federal grants, however, tend to provide for single projects, within a limited time-frame, and may lose some of this efficiency.

Other sources of funding in addition to federal grants are available for the building and operation of communications systems. Given the many potential sources of funds available to states and communities, it is debatable whether rules attached to federal grants can be used to drive a process that would coordinate the construction and operation of a seamless, nationwide network to serve first responders and other emergency personnel.

Decision Making and Planning Authority

As with federal grant programs for emergency communications, multiple federal agencies are involved in planning and decision-making guidance for state, local, tribal, and regional agencies. Although a number of federal programs provide support for state and local activities, there appears to be little coordination among these agencies in carrying out their separate programs.[30]

Federal Involvement

Congress has separately and specifically given authority to DHS and to the FCC to act on behalf of public safety. In the case of DHS, authority includes planning and implementing public safety communications solutions and setting requirements to coordinate and support specific goals, such as interoperability and a national communications capability.[31] Within DHS, the Office of Emergency Communications manages statewide planning and coordination for interoperable communications and administers compliance with the National Emergency Communications Plan.[32] FEMA participates in emergency communications planning as part of its state and regional efforts for hazard mitigation and response and recovery.[33]

The FCC has several pertinent mandates from Congress, such as an obligation to "promote safety of life and property through the use of wire and radio communication,"[34] and requirements regarding the assignment of radio frequencies for public safety use. The FCC created a Public Safety and Homeland Security Bureau in 2006 to consolidate its many programs oriented toward public safety.[35] This bureau and the Wireless Telecommunications Bureau jointly administer the Public Safety Broadband Licensee, the chosen license-holder for the public safety broadband frequencies at 700 MHz.[36] The FCC also oversees the activities of 55 Regional Planning Committees that administer spectrum assignments for mission-critical voice networks in the 700 MHz band.[37]

Other federal agencies with decision-making responsibilities for public safety planning and investments include the NTIA and the National Institute of Standards and Technology (NIST). For example, the Public Safety Communications Research program is a joint effort between NIST's Office of Law Enforcement Standards and the NTIA.[38]

State and Local Involvement

Every state has one or more agencies that plan for public safety, homeland security, and emergency communications. Most states have a Statewide Interoperability Coordinator (SWIC)[39] to administer its Statewide Communication Interoperability Plan (SCIP). SCIPs are written to conform with federal guidelines and requirements, such as the NECP, and provide guidelines and goals. DHS describes SCIPs as "locally-driven, multi-jurisdictional, and multi-disciplinary statewide plans to enhance emergency communications."[40] State coordinators are encouraged to promote a "coordinated practitioner-driven approach" through a "collaborative statewide governance structure."[41] Not all grants are coordinated through the SCIP, however. Many earmarks and some grants have been awarded directly to an agency within a town, city, or county for a specific purpose identified by the recipient. A review of federal, state, and local grants programs seems to indicate that most purchasing decisions for emergency communications equipment occur at the county[42] level, often based on an approved equipment list provided by the state or by a regional buyers cooperative.[43]

Planning and decision making for emergency communications at the local and county level is often separated by function. Some actions may be coordinated through a planning board, or one office, such as the sheriff's office, will be designated for emergency planning. Federal and state programs may provide specific goals and a framework for coordination. It appears, however, that funds are disbursed and authority is typically dispersed according to the administrative structure of the county, city, or town. There is a great deal of variation across the country but, generally, state officials manage communications for statewide systems, such as the state highway patrol, and local officials manage local public safety services.

Commercial Involvement

In its *National Broadband Plan* and other documents, the FCC has advocated some form of public-private partnership between public safety agencies and wireless carriers that would include sharing responsibility for planning and decision making. The decision by the FCC to advocate a shared network for wireless communications was centered on two conclusions: (1) that a network with national coverage would best meet public safety needs for robust communications capabilities, information, and interoperability; and (2) that sharing spectrum and network facilities with commercial users would benefit public safety through economies of scale in building, equipping, and operating the network, by providing access to additional spectrum in times of large-scale emergencies, and by offering new sources of funding, among other benefits. Arguments in favor of building a network exclusively for public safety revolve around the shortcomings of current commercial wireless services such as poor availability, inadequate coverage in rural areas, lack of security features, and absence of priority access. Also, public safety representatives have frequently expressed concerns in testimony before Congress about sharing authority with a commercial partner.[44]

Final decisions by policy makers about the number, location, and type of network (local, state, regional, national) are likely to have a significant impact on commercial participation in a public safety broadband network or networks. Although most public safety representatives object to any sharing arrangement that would allow commercial partners to dictate access to capacity, in general they have advocated leveraging public safety spectrum resources through agreements with commercial partners. The scope of these agreements is undecided. Some agencies envision a national network governed primarily by the Public Safety Broadband Licensee; some prefer linking state and regional networks, with regional boards taking the lead; others assume thesolution will be a system of multiple local, county, and state systems similar to what is in place for existing public safety radio systems.[45]

To maximize the benefits of commercial involvement, the National Emergency Number Association (NENA) in 2009 proposed consolidating the Public Safety Broadband License with the D Block, to be auctioned to commercial operators who would be required to develop public safety enterprises to provide the needed network.[46]

Role of the Emergency Response Interoperability Center and the Public Safety Spectrum Trust

As part of its initial planning for a Public Safety Broadband Network (PSBN),[47] to facilitate interaction between the public and private sectors, the FCC required the creation of a not-forprofit corporation, the Public Safety Broadband Licensee (PSBL) to hold the license for spectrum assigned for public safety broadband. This corporation is presently governed by its board, the composition of which is largely determined by the FCC, with the participation and oversight of two FCC bureaus. As originally envisioned by the FCC, the PSBL would be contractually bound to work with one or more commercial partners to develop a broadband network for public safety use. The FCC awarded the administration of the PSBL to the Public Safety Spectrum Trust (PSST).[48] To work with the PSBL, the commercial sector, and other stakeholders, the FCC created the Emergency Response Interoperability Center (ERIC) within the FCC Public Safety and Homeland Security Bureau in April 2010.[49] ERIC was tasked with implementing standards for national interoperability and developing technical and operational procedures for the public safety wireless broadband network in the 700 MHz band.[50]

The national license currently assigned to the PSST for a broadband network is for 10 MHz within the 700 MHz band.[51] Until plans are finalized for the PSBN, the FCC has provided conditional waivers for some agencies that wish to move forward with broadband network deployment. Agencies in eight states, five counties or regions, and eight cities received waivers, which required them to meet network requirements as specified by the FCC. The FCC then approved agreements for 20 of the waiver recipients to lease spectrum from the PSST to cover their planned areas of operation.[52] Transferring spectrum management rights to these jurisdictions effectively assigned much of the decision-making authority and funding responsibility to them as well.

Based on FCC plans, however, ERIC and the PSST will do most of the planning for the PSBN; future grants intended for the PSBN would have to be used in accordance with the network plans developed by the FCC through ERIC. The FCC, therefore, would be the lead agency for grants guidance, governance, planning, and some decision making for the PSBN. The level of decision-making power that the FCC will be able to wield depends largely on the outcome of its negotiations with the PSST and other public safety stakeholders, possibly bolstered or constrained by congressional action.

Given the apparent expectations of many stakeholders that there will be commercial participation in building and operating a nationwide network for public safety broadband communications, it might be useful to develop specific guidelines for these partnerships at the beginning of the network planning process. For example, it is not clear whether the PSBL will take the lead in negotiations and contractual agreements with commercial partners or whether decisions regarding these agreements will be left largely to states, or localities, as has been the case in the past.

Although the FCC would appear to have authority over the process of planning and decision making for the PSBN, it may come under pressure from stakeholders, including Congress, to share responsibilities. Collaborative decision making among a large number of stakeholders may dilute efforts for an efficient, cost-effective, interoperable network, some say, without necessarily representing the interests—or meeting the needs—of its users.

Cost Estimates

Building the network requires capital expenditures for infrastructure and operations centers, wireless towers, radios, and other communications equipment. Covering on-going expenses requires additional funding for software upgrades, maintenance, network operations, training to use new equipment, and other recurring costs. For wireless communications, an important infrastructure component is the network that links radio towers to communications backbones. These networks, which usually operate over fiber-optic cable or microwave connections, are typically referred to as backhaul.

In 1996, the Public Safety Wireless Advisory Committee reported that the estimated value of the installed base of nonfederal public safety communications networks was $25 billion, with a useful life of 15 years.[53] Most of the networks in place at the time operated on analog frequencies.

In recent years, the National Emergency Management Association (NEMA) has reported on interoperable communications expenditures in at least two of its biennial reports.[54] These reports provide summaries of responses to questionnaires sent to NEMA's members. In 2006, NEMA reported that states estimated that it would cost a total of $7 billion to achieve state-wide interoperability for national coverage or to reach levels required by each state's homeland security strategy. The average expenditure, per state, for states providing estimates of their projected costs, was reported to be $160 million.[55] In 2008, NEMA reported that the states' estimates of the cost of providing interoperable communications nationwide had risen to a total of $12 billion since the 2006 report. Obstacles to achieving interoperability that were cited in the report include rapidly changing technologies that require repeated and costly investments; lack of cooperation among jurisdictions; and a lack of expertise and resources for proper planning.[56] The 2010 Biennial Report carried no information about interoperable communications.

Wireless Networks

In 2008, when the FCC proposed a public-private partnership to build a public safety broadband network, the cost was estimated in public comments filed with the FCC at from $18 billion to as much as $40 billion.[57] These projected costs did not include radios. The network would have been largely built by the purchaser of a commercial spectrum license, known as the D Block, allocated for that purpose. The planned auction of the D Block failed.[58]

In 2010, the FCC again proposed to auction the D Block, under new rules that would include sharing infrastructure with a separately funded and managed Public Safety Broadband Network.[59] Although the FCC did not provide details of how the D Block would be auctioned, it did prepare estimates of the cost to public safety if infrastructure was shared. These estimates were based on assumptions that include (1) 95% of the towers used in the PSBN would be shared with commercial operators; (2) backhaul would be shared and paid for separately; (3) the new network would be data and video only; and (4) LTE would be the wireless technology used. The FCC concluded that the "total present value of the capital expenses and ongoing costs for the network over the next ten years is approximately $12-16 billion." Of this, $6.5 billion would be for capital expenditures. The cost of radios is not included.[60] The FCC also provided an estimate for building a stand-alone public safety network. Instead of $6.3 billion in capital expenditures for building and equipping tower sites, a stand-alone network would cost $12.6 billion. Additional investments of $3.1 billion would be required for backhaul and operations centers, bringing the total estimated investment to $15.7 billion.[61] The total present value for both capital expenses and operating costs over 10 years for a stand-alone network might be $34.4 billion to $47.5 billion.[62] Both the incentive-based and stand-alone network cost estimates are based on 44,800 cell sites. Costs might be reduced if pre-existing public safety sites for cellular network connections can be used and might be increased if networks are operated by multiple jurisdictions, leading to inefficient redundancies.

Network Size Impacts Cost

Communications capabilities generally available for the emergency response community tend to be more costly than commercial solutions. Although some of the added cost can be attributed to unique requirements such as operability in extreme weather conditions, much of the higher cost comes from limited production runs in an industry otherwise characterized by aggressive policies to achieve economies of scale.

To take advantage of economies of scale and provide efficiency in investment, operating costs, and spectrum use, among other benefits, the Governing Board of the National Public Safety Telecommunications Council (NPSTC)[63] has recommended the adoption of a national network with a small number of core networks for public safety broadband.[64] In its conceptual model, NPSTC has placed seven distributed core networks within the continental United States to provide national coverage and necessary system redundancy.[65] NPSTC has stated that this national network architecture approach would be less costly than any of the "network of networks" architectures that have been proposed by numerous public safety officials. According to NPSTC,[66] the cost of the extra capital needed to build a network of networks would be "unnecessarily expensive." One nationwide network "will cost the least amount to manage, maintain, and upgrade." NPSTC also observed that "recurring costs to manage and refresh a network, over the long term, will be the largest cost component regardless of configuration." According to NPSTC, coordinating the funding and installation of system upgrades across multiple networks would be difficult and inconsistencies could lead to operability and interoperability problems.

Network Capacity Impacts Cost

A cost analysis based on research done at Carnegie Mellon University provides alternative scenarios using different assumptions about the type of public safety communications network to be built.[67] For example, three scenarios are considered for a public-safety-only network built on 10 MHz of spectrum and for a public-private-partnership network using 20 MHz of spectrum. The three operational scenarios are (1) voice only; (2) data only; and (3) voice and data.

The estimates are derived from assumptions about network traffic, capacity, and technical parameters that lead to an estimate of the number of cell sites needed for each type of network. Costs are estimated based on the cost of building and operating the expected number of cell sites needed for each scenario. Deployment costs are those associated with the installation and operation of cell sites and do not include other network components such as mobile switching centers. The estimates use values of $500,000 per cell site for deployment and $75,000 per cell site, annually, for operating costs. The cost of buying, maintaining, or upgrading radios is not included. Also, the estimates for a public-safety-only network are based on 10 MHz of available spectrum; no analysis was done for a public-safety-only network using 20 MHz. The cost analysis yielded an estimated investment of $11.1 billion to deploy a voice and data network on 10 MHz of public safety spectrum. The comparable estimate for a public-private network on 20 MHz is $9.7 billion. The White House has, in some documents, assumed an investment of $10 billion is needed to deploy a wireless broadband network for public safety using 20 MHz of dedicated spectrum.[68]

Table 1. Network Scenarios and Estimated Costs (In millions of dollars)

Network Scenario	Public Safety Only	Public-Private-Partnership
Voice Traffic Only		
Total Number of Cell Sites	3,700	10,300
Upfront Deployment Cost in Millions	$1,900	$5,200
Operating Cost in Millions/Year	$280	$770
10 Year Total Cost in Millions	$3,400	$9,400
Data Traffic Only		
Total Number of Cell Sites	18,200	19,000
Upfront Deployment Cost in Millions	$9,100	$9,500
Operating Cost in Millions/Year	$1,400	$1,400
10 Year Total Cost in Millions	$16,600	$17,300
Voice and Data Traffic		
Total Number of Cell Sites	22,200	19,400
Upfront Deployment Cost in Millions	$11,100	$9,700
Operating Cost in Millions/Year	$1,700	$1,500
10 Year Total Cost in Millions	$20,300	$17,700

Source "Quantifying the Costs of a Nationwide Public Safety Wireless Network" by Ryan Hallahan and Jon M. Peha, authors' version.

Network Infrastructure: Wireless Backhaul and NG9-1-1

Wireless communications over the airwaves come to earth through towers and switches that link them to telecommunications and Internet networks. As telecommunications technology evolves from voice traffic over the Public Switched Telephone Network (PSTN) to data traffic over the Internet and Internet-like networks, the nature of network support for wireless traffic is also changing. To take advantage of IP-enabled technology, some states and communities have built their own links to backbone networks to supplement or replace network access through commercial carriers. Other communities have built their own networks because commercial carriers were not providing service to their area. Where these networks are in place or are being developed, IP-enabled public safety technologies are often making the transition to these networks. The networks can be designed to support NG9-1-1, emergency alerts, and wireless network backhaul, as well as the broadband communications needs of government agencies and the general public.

Studies published by the Department of Transportation in 2008 estimated the cost of providing 911 services for several scenarios over a life cycle extending from 2009 through 2028. One group of assumptions would maintain the existing 911 infrastructure. Cost estimates for two different scenarios supporting that assumption set cumulative investments in "acquisition and implementation" at $9.3 billion and $13.1 billion, depending on the assumptions and the weight given to them. Similarly, cumulative "operations and maintenance" costs were estimated at $46.4 billion and $65.8 billion.

Another set of assumptions was developed to estimate costs for building a new NG9-1-1 network as a stand-alone network. The estimates were $8.7 billion and $9.1 billion in cumulative investment, and $51.1 billion and $49.2 billion in operating costs.[69] The estimates for new, NG9-1-1 systems do not take into account how much those costs might be reduced by sharing infrastructure. Most states, however, do plan on sharing network infrastructure among multiple users. A number of BTOP grants, for example, have been made for the construction of multipurpose broadband networks that include 911 services and backhaul for first responder radios.

Radios

In addition to cooperation for sharing network resources, the FCC has anticipated that commercial partners would lead, and fund, the development costs of the radio technologies that will operate within the frequencies assigned to the Public Safety Broadband License and the D Block. The participation of commercial carriers in developing and deploying crucial radio components is expected to put the cost of public safety radios in the same price range as commercial high-end mobile devices ($500). By contrast, interoperable radios for the narrowband networks at 700 MHz cost $3,000 and up, each. The latest radios developed for public safety by DHS, designed to operate on multiple bands, are estimated to cost between $4,000 to $6,000.[70]

Among the many unresolved debates surrounding the development of a broadband wireless network for public safety use is the feasibility of expecting the new network to carry both data and voice traffic. Commercial LTE networks will in the immediate future use existing voice-data networks (primarily third-generation networks, 3G) to provide the voice component for LTE (4G) devices. This is not considered a viable option by most public safety officials, as the 3G networks do not support a number of features considered essential for public safety voice communications. Many of these features are dependent on radio-to-radio (off-network) communications that today require assigning channels for exclusive use. Some radio communications experts argue that LTEbased solutions for public safety's needs can be developed and adapted rapidly. Others believe that the process of integrating voice and data will be difficult and slow, taking perhaps decades. Some public safety officials believe it is not possible for all their needs to be met by a single, unified network and that separate radios and networks for separate functions are the solution.[71]

Competition and Cost

The FCC has concluded that proprietary technologies have hampered the effective development of public safety radios and curtailed interoperability, based on findings provided to Congress in response to an inquiry from the House Committee on Energy and Commerce in 2010.[72] In April 2011, the committee wrote to the FCC with a list of questions about the selection of equipment vendors by those agencies that have received waivers to begin building parts of the PSBN. Questions related to policy issues raised in this report include the possible impact of proprietary broadband technologies on

> network and device equipment costs borne by public safety relative to commercial costs; the ability for public safety to benefit from innovation in wireless technologies;
>
> the likelihood of terminated product lines or new mandatory releases that result in unique costs to public safety relative to commercial technologies;
>
> public safety interoperability at the application, devices, and network levels among networks provisioned by different vendors;

the ability of public safety users to enter into partnerships with commercial wireless providers;

competition in the public safety communications equipment market; andthe FCC's National Broadband Plan finding that encouraged incentive-based partnerships with a variety of commercial operators.[73]

The FCC responded to this letter in mostly general terms, citing, for example that the waiver process is in its early stages.[74] The FCC response noted that issues similar to those raised by the committee letter are being examined in the *Interoperability Further Notice of Proposed Rulemaking*.[75] The FCC's primary focus has been to develop rules to ensure technical interoperability nationwide for the public safety broadband network, even if proprietary technologies are used. The FCC reply noted that, of the 20 waiver recipients, only two had selected a vendor, and another eight were in the process of identifying and selecting a vendor. The FCC does not require waiver recipients to use competitive bidding because it does not oversee the procurement practices of state and local governments.

Network Access

The FCC has based its planning and cost projections on the assumption that there will be a nationwide network administered by a single license-holder. Meanwhile, states and localities have continued to apply to the FCC for waivers that would permit them to obtain sub-licenses from the PSST and commence building their own networks. Any IP-enabled network, however, can support virtual private networks in wired or wireless infrastructure. The most important consideration for assured access, therefore, is not network ownership but network availability, permitting the uninterrupted transfer of communications to an operational link when part of a network fails. Sharing infrastructure between public safety and commercial license-holders, as proposed by the FCC, would lower costs and increase national availability by providing access to commercial networks. A nationwide public safety network, with a small number of operational cores operated by and for public safety agencies, as proposed by NPSTC, would increase network availability and facilitate cost-saving measures for participating public safety agencies. Turning over network construction and operations to the commercial sector, as NENA has proposed, might lower costs even further but might also limit public safety access, unless strict rules for meeting capacity needs are in place. If multiple jurisdictions build their own public safety broadband networks, not only might the total cost of providing capacity increase significantly but also available capacity might be diminished because of the added complexity of coordinating interoperable access. The choice of design for the network, therefore, may have an impact on the amount of funding and the length of time needed to construct an interoperable network that can be accessed nationwide.

Once governance and decision-making authority can be conclusively decided, it may be possible to provide the nation not just with a plan, but with a network design that provides a more solid basis for cost estimates and for plans for funding. States and localities, for example, might benefit from a better sense of their future financial obligations in the construction and operation of a broadband network that is to provide access to public safety agencies across the country.

Public safety agencies have multiple obligations to build or upgrade, and equip, other networks. The costs of building and maintaining a new data network, therefore, are some fraction of total obligations to assure emergency communications capabilities. Based on information about the cost of existing systems and estimates of future costs, the construction of this new network represents a significant investment for all participants.

Financial Resources

Although there are many ways that the federal government might encourage improvements in emergency communications, such as tax incentives or cash awards in technology contests, most federal financial support has come from grants and congressionally directed funding. Other sources of funding for emergency communications include appropriations from state and local budgets, financing from government bonds, grants from private foundations, lease-purchase agreements with equipment suppliers, and sources of recurring revenue such as fines, user fees, surcharges, and state and local sales and property taxes. At the local level, funds are often collected through bake sales, fish fries, fund-raising drives, and other community-supported efforts.

Testimony at congressional hearings and other public statements by public safety officials indicate that many public safety agencies envisage setting up a partnership with commercial network operators that would provide some form of revenue. How this would be accomplished has been left vague. Possible models to generate revenue from public safety spectrum assets include issuing secondary licenses to commercial partners, and network management techniques that use advanced technologies to share spectrum.

The final decision as to how spectrum access might be divided up across an estimated 65,000 public safety agencies[76] has not yet been made by the stakeholders. Some municipalities and states may seek commercial partnerships that will provide access to public safety spectrum in return for financial consideration. Resources provided by the commercial sector might include access to their infrastructure and cash payments for spectrum leases. The value to commercial operators of spectrum access would likely depend on the geographical coverage being offered by the public safety licensee. Paradoxically, the areas where the spectrum is likely to have the most value, such as urban areas, are also the areas where public safety agencies are most likely to need spectrum capacity for their own use. Leased or shared spectrum in small or isolated communities may have little commercial value.

The monetary value of spectrum licenses might also be captured for public safety communications expenses by using the proceeds from commercial license auctions. Several proposals, including proposed legislation,[77] would designate all or part of the proceeds of some auctions to funding public safety communications investments and operating costs. These proceeds would be deposited in a special fund; grants administrators would be allowed to borrow against anticipated future revenue so that grants could be provided immediately.

Congress has twice in the past acted to create special funds to receive and distribute revenue from spectrum auctions for specific purposes. These funds represent a departure from existing law, which requires that auction proceeds be credited directly to the Treasury as income.[78] The Deficit Reduction Act of 2005 (P.L. 109-171, Title III) required the auctioning of licenses for spectrum used by television broadcasters for analog transmissions. It established the Digital Television Transition and Public Safety Fund to receive this auction revenue and use some of the proceeds for the transition to digital television, public safety communications, and other programs. The Commercial Spectrum Enhancement Act (P.L. 108-494, Title II) established a Spectrum Relocation Fund to hold the proceeds of certain spectrum auctions for the specific purpose of reimbursing federal entities for the costs of moving to new frequency assignments.

Current law prohibits the FCC from considering potential spectrum revenue in planning auctions of spectrum licenses,[79] yet high auction yields may be necessary to sustain the proposed grant programs. Furthermore, although auctions of spectrum licenses are an effective solution for implementing today's wireless technologies, spectrum management policies might change to accommodate the technologies of tomorrow, with less reliance on auctions as a policy tool. It may be that additional congressional oversight is required to assure that the maximum national value is obtained from national spectrum assets.

Using revenue generated by the sale of radio frequency spectrum to fund wireless networks might increase the proportion of federal money available for one-time investments in infrastructure and therefore the federal role in decision making. This, however, might require transferring authority from states and communities to federal agencies, leading to a greater level of federal participation than is currently the case. States and localities might be hard pressed to muster the resources needed to participate. Additional funding sources, such as private sector investment, may need to be considered as plans for the proposed network move forward, some say.

Managing Radio Frequency Spectrum

Decisions about the assignment and management of the spectrum resources available to public safety agencies often provide the boundaries for making other decisions, such as choices for technology, governance, and funding. For example, within the 700 MHz band, 24 MHz has been assigned for public safety use, of which 10 MHz is currently designated for a new broadband network.[80] Networks primarily for mission-critical voice communications are being constructed separately on 12 MHz of public safety's allotment within the 700 MHz band. Construction of these voice networks, referred to as narrowband networks, is being coordinated largely between states and counties, with counties and major cities typically taking responsibility for building on the spectrum assigned to them for that purpose.

Figure 1 represents part of the upper 700 MHz band, indicating the frequencies assigned for public safety and adjacent commercial licenses. The accompanying legend provides the key to the type of license and the amount of spectrum associated with that spectrum. For example, reading from the left, the first band represents 11 MHz of the C Block, which is for commercial purposes. The C Block allocation is in two 11-MHz assignments, at 746-757 MHz and at 776-787MHz.

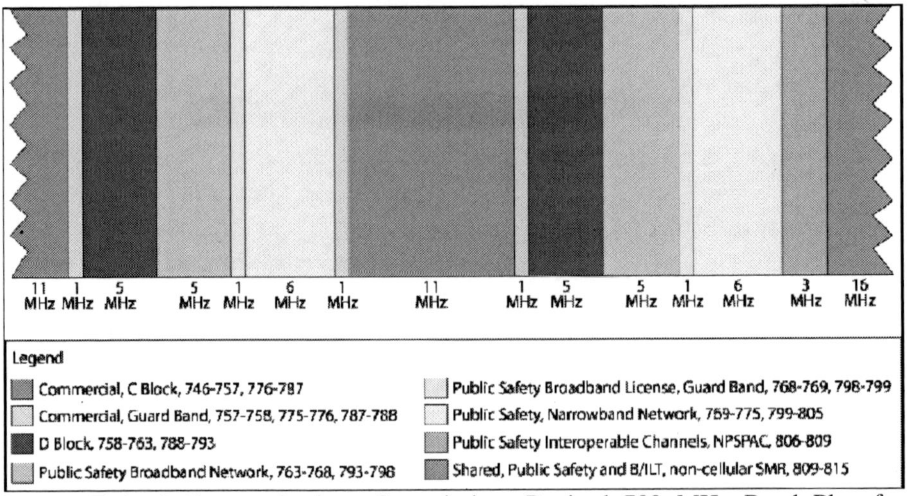

Source: Federal Communications Commission, Revised 700 MHz Band Plan for Commercial Services.

Figure 1. Public Safety and the D Block.

The D Block has two assignments of 5 MHz each; these are contiguous with the two assignments for the Public Safety Broadband License. The D Block frequencies have not yet been auctioned orotherwise assigned. Assigning the 10 MHz D Block for the Public Safety Broadband Network would increase the broadband network's coverage to 20 MHz. Not all spectrum blocks in the 700 MHz band are shown in *Figure 1;* other comer-cial licenses are in Bands A and B in the lower part of the band.

Public safety officials, commercial network experts, and the FCC are generally in agreement that LTE technology should be required for the new PSBN. Part of the challenge for network developers will be to coordinate the development of public safety requirements for LTE with commercial LTE standards development.

Coordinating development of 700 MHz band standards among network participants provides an opportunity to maximize the benefits inherent in IP-enabled networks for the safety of the general public. For example, it is possible to create smart phone applications that can link personal mobile devices to emergency command centers, integrating information from those devices into an action plan for response and recovery. The *first* first responders—the people at the site of a disaster—can be enabled to participate proactively in the response. The feasibility of using crowd-sourced communications to provide information to emergency management command centers is being tested by a program undertaken jointly by the Los Angeles County Homeland Security Advisory Council and the Annenberg Innovation Lab:[81] the CrisisConnection Project.

The CrisisConnection Project, as described by Professor Gabriel Kahn, a faculty associate at the Innovation Lab who is leading the development team,[82] will encourage businesses and others inthe Los Angeles area to download a smart phone app or register a cell phone number.

When a disaster strikes, a user can snap a photo of damage, such as a downed power line from a storm or a collapsed building in an earthquake. A geo-location function then pinpoints where and when the photo was taken. That information is uploaded onto a network. A mapping function places the photo on a map accessible from the web. Other information, such as the names of people who were able to exit a building before it collapsed or whether a city street is still receiving its water supply, can also be uploaded. All the information is aggregated onto one map.

Fully implemented, the data might help emergency situation managers determine where to most effectively deploy emergency medical service personnel, firefighters, HazMat teams, utility repair crews, or other response and recovery personnel, as appropriate.

As a situation stabilizes, evaluations about evacuation routes, shelters, and other post-disaster services could be expedited and information disseminated through emergency alert systems.

The site can be used to support local 911 call center operators and link to NG9-1-1 networks for broader coverage.

The scope of such life-saving measures would be greatly facilitated by effective network-tonetwork wireless communications. For example, federal policy regarding access to the LTE networks being built in the 700 MHz band might include requirements to leverage these spectrum assets for a nationwide emergency communications grid that meets the safety needs of the public at large.

One the goals of effective spectrum management is to create opportunities for the development of innovative technologies. Wireless technology transforms air into desirable services, providing an engine for economic growth and development. The evolution of public safety communications has lagged behind the commercial sector and the military in receiving the benefits of recent innovations.

By providing access to desirable spectrum in the 700 MHz for public safety purposes, Congress has provided an opportunity not only for increased network capacity but also for increased innovation in emergency communications technologies.

Technology

Separately, public safety agencies across the country are investing in public safety communications infrastructure for other technologies on other frequency assignments. In addition to the proposed LTE data network at 700 MHz, there are other networks operated by public safety agencies on dedicated spectrum: narrowband networks on frequencies below 512 MHz; a separate narrowband network using spectrum at 700-800 MHz; and ultra-wideband, short-range networks at 4.9 GHz.[83] Public safety also uses unlicensed spectrum for local voice and data networks. These networks and other communications solutions have been built separately, use different technologies, and support different radios.

The commercial sector, meanwhile, has begun the transition to operating almost exclusively on IP-enabled networks such as LTE. Wireless carriers around the world are installing LTE networks for consumers and planning for the next generation of LTE: LTE Advanced.[84] LTE Advanced technologies will be able to operate across noncontiguous spectrum bands, thereby increasing channel widths for greater capacity and performance. Most experts agree that LTE Advanced will facilitate the transition to new technologies by making it easier and less expensive to phase out older infrastructure.

Many experts in advanced communications technology believe that the push to IP-enabled technologies is likely to bring about the convergence of commercial, military, and emergency response (federal and nonfederal) technologies on common, interoperable platforms. In this view, compatible communications devices will be differentiated by applications developed by stakeholders to meet their mission needs. Infrastructure, spectrum, and mobile devices will be sharable, and it is envisaged that sharing will be encouraged.

Networks

The military is linking many of its communications through IP-enabled networks,[85] similar to plans by the public safety community for investment in first responder LTE devices and NG9-1-1. The Department of Defense (DOD) has used the term internetwork to refer to the IP-enabled networks that drive its Global Information Grid (GIG) for network communications.[86] The internetwork, also known as the Convergence Layer, provides analysis and organization of communications activity to facilitate transport. The communications layer that provides the entry and distribution links to services supported by the internetwork is referred to as the Link and Transport Layer by DOD.

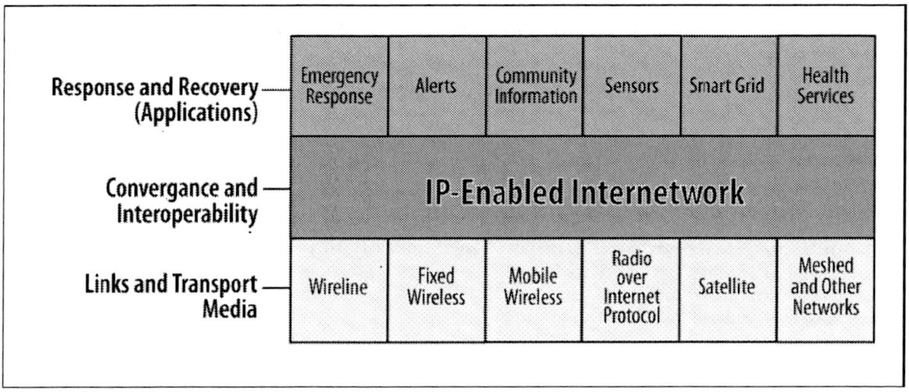

Source: Based on Department of Defense Global Information Grid Architectural Vision; vision for a net-centric, service-oriented, DoD enterprise, prepared by the DOD CIO, June 2007.

Figure 2. Emergency Communications Grid.

An Emergency Communications Grid, similar to the military's GIG, might use a common IP-enabled network structure to connect with any type of IP-enabled system, network, or device to support a wide range of services. (See Figure 2.)

DOD's internetwork is the equivalent of public safety wireless backhaul, NG9-1-1 network connectivity, or any other network connection that serves the public interest. The EmergencyCommunications Grid shown in Figure 2 can also send out emergency alerts to endangered populations,87 capture information from sensors, manage the Smart Grid to prevent power outages, and support other response and recovery actions. In nonemergency mode, the shared internetwork serves as the conduit for the daily workload of communications.

The Emergency Communications Grid represents a unified approach to assuring access and interoperability among all types of communications devices and applications, but it is not envisioned as a single network. The internetwork would be a convergence of many IP-enabled networks that support all the necessary devices and provide the needed links to the Response and Recovery level.

Radios

The United States Army is adapting commercial wireless technologies to operate with military networks, expanding the range of available devices and applications.[88] It has initiated a project, Multi-Access Cellular Extension (MACE), to capture innovation in the commercial marketplace because the "commercial marketplace continuously introduces new technologies, replete with greater capabilities and faster data rates, which could prove beneficial to the military." The vision for MACE is intended to unify commercial technologies with military needs.

The developers of MACE contend that it will support cellular communications over smart phones as well as links through cellular base stations to tactical systems. In demonstration pilots, mesh networking[89] is to be used to test seamless operation in WiFi/cellular environments even when a cellular base station is not available.[90]

One example of a smart phone app being tested in a MACE demonstration is the mapping of Improvised Explosive Device locations. Similar to the CrisisConnection project described in the preceding section, the smart phone camera would take a picture at the site of an explosion and, using location technology, pinpoint it on a map of the area. This information would be used in planning routes to be taken through the area.[91]

The MACE strategy, and other Army Research and Technology initiatives that are augmenting commercial solutions to work in the military environment, may also be applied for public safety response and recovery efforts. The developers of MACE have identified a number of parallels between tactical military deployments and first responder deployments. For example, militarydivisions and public safety personnel both need robust communications technology to provide: situational awareness, chain-of-command and network management functions, authentication and access control, and data security—at a low cost.

Migration to Commercial Technologies

The military model may not be readily transferable to public safety, however. The approach being taken by the U.S. military is to set up structures that incorporate commercial technologies and innovation into existing tactical systems. Public safety communications do not have a similar system on which to build.

The new narrowband and broadband networks for public safety will, according to testimony, be incompatible with each other and with other networks for the foreseeable future. Only a small part of the existing public safety infrastructure is expected to be usable in the development of new networks at 700 MHz. The military is planning to use commercial technologies to expand the capabilities of its existing communications base, public safety agencies, on the other hand, may need to use commercial infrastructure as well as technologies. Using commercial infrastructure to support public safety apps would emulate MACE's plan for melding commercial and military technologies to capture the benefits from innovation and cost-efficiencies. There is a generally recognized need to move to technologies that use Internet protocols in order to improve service and reduce operating costs. Because IP-enabled technologies are radically different from those of most of the emergency communications systems currently in place, the transition provides opportunities to surmount existing barriers to change. Just as access to the Internet has revolutionized business and social cultures worldwide, access to IP-enabled networks is likely to reshape the command-and-control hierarchy of first responders, and the ways they use and pay for communications systems.

Public safety communications may be closed off from vital new technologies unless solutions are found to incorporate and foster change. Communications technology for public safety, as for the military, needs to be regularly refreshed by innovations and cost-saving efficiencies developed by a competitive commercial sector, according to some experts.

LEGISLATION IN THE 112TH CONGRESS

The House of Representatives, on December 13, 2011, approved the Middle Class Tax Relief and Job Creation Act of 2011 (H.R. 3630, Representative Camp). H.R. 3630, Title IV contains provisions from the Discussion Draft of the Jumpstarting Opportunity with Broadband Spectrum (JOBS) Act of 2011, as amended, approved in markup on December 1, 2011, by the Subcommittee on Communications and Technology, House Committee on Energy and Commerce. H.R. 3630 is under consideration by a conference committee for which a major focus of discussion is an extension of payroll tax cuts and how to fund them. In the Senate, the legislative response to H.R. 3630 may include provisions from the American Jobs Act of 2011 (S. 1549 and others) and from the Public Safety and Wireless Innovation Act (S. 911, as amended, Senator Rockefeller). S. 911 received bi-partisan approval by the Committee on Commerce, Science, and Transportation.

This section provides an overview of H.R. 3630, Title IV and what are referred to as the "administration bills" because they reflect administration policies on public safety and spectrum policy. These are: S. 911, as amended; H.R. 2482; and Title II, Subtitle H of S. 1549, and similar bills S. 1660 and H.R. 12. The House bill references are the pertinent sections of H.R. 3630. Specific citations for the administration bills are from H.R. 2482, the companion bill to S. 911, as amended, and S. 1549.[92]

There are a number of possible paths forward for federal policy. One approach might be to return to the pre-9/11 era, when the federal government left most planning and purchasing decisions to state and local governments, and federal and nonfederal network managers negotiated reciprocity agreements, separately and independent of each other. Since 9/11, however, the technologies that might provide effective and efficient communications for public safety needs have evolved more quickly than the organizational structures of the agencies that plan for, use, and manage these technologies.

The rapid transition to broadband communications, for example, has widened the gap between practice and expectation; how public safety agencies have managed their communications assets in the past may not be a good model for how they should be managed in the future. The administration bills currently before Congress would provide government support to move public safety agencies away from the incremental development of narrowband voice networks to next-generation IP-enabled networks that carry voice, data, and video transmissions. They would assign the D Block and Public Safety Broadband licenses to a not-for-profit Public Safety Broadband Corporation, and would fund its initial costs to build a single, nationwide network. Additional funding would be from loans and fee income.

H.R. 3630 would provide federal leadership to develop a plan that ensured interoperable broadband communications among states. A federally appointed Administrator would ensure compliance with the requirements and manage distribution to states of frequency assignments in the D Block and Public Safety Broadband licenses. States that participated in the plan would assume the primary responsibility for building and operating individual state networks in partnership with the commercial sector.

The federal government would provide funds for a matching grant program. States would determine how to fund their own networks but would be required to work with commercial partners and would be allowed to enter into partnerships that would generate fee income from the commercial sector.

Provisions for Public Safety Broadband Network Deployment

Planning and implementation of broadband networks for public safety communications are still in the early stages. Among the barriers to moving forward are incomplete development of technology and standards; inadequate planning; insufficient coordination among public safety agencies; lack of governance structure to direct and administer a nationwide, interoperable network; and lack of sufficient funding. The bills would provide mechanisms for planning and governance and would establish grants programs for financial assistance. The administration bills would create a not-for-profit corporation to undertake the construction and operation of a nationwide broadband network. H.R. 3630 (§4203) would create an Administrator and a Public Safety Communications Planning Board (§4202) to direct the actions of individual states to carry out the construction and operation of state systems. States would not be obliged to build a network, but would be required to adhere to the requirements established through the act if using the spectrum designated for public safety broadband communications. Local networks within states would be allowed, as would regional networks across states, as long as these were compliant with the act.

Public Safety Spectrum Assignment and Use

All three bills would require the FCC to assign an additional 10 MHz of radio frequency spectrum in the 700 MHz band for public safety use. These frequencies—at 758-763 MHz and 788-793 MHz—are known as the D Block. The bills would combine the existing Public Safety Broadband License (currently assigned to the Public Safety Spectrum Trust – PSST) and the D Block in a single license. The FCC would be required to take the steps necessary to transfer the license.

The administration bills would assign the combined license to a newly created Public Safety Broadband Corporation. H.R. 3630 (§4201) would assign the license to the Administrator of the National Public Safety Communications Plan, as required by the bill.

The administration bills would specify that the broadband license be administered as a single, national license. H.R. 3630 would require the Administrator to oversee the use of the licensed frequencies by each state, or regional grouping of states, exclusively for broadband networks, as described in the bill.

Public safety agencies are also building separate networks within the 700 MHz band[93] that are referred to as narrowband because they support voice communications but do not have sufficient bandwidth to handle complex data or video. These narrowband networks are being built primarily by counties or urban areas although many of the networks are intended to be statewide.

The administration bills would affirm the FCC's right to allow flexible use on these narrowband frequencies, subject to conditions. Both bills would require the FCC to prepare reports on the efficient use of spectrum but with markedly different requirements for the contents of the report. S. 1549 would require the FCC to focus on improving the efficiency of spectrum used by public safety agencies, and in particular to assess the feasibility of relocating from other bands to the Public Safety Broadband Network. Vacated spectrum might be reassigned through auctions, including incentive auctions. H.R. 2482 would require the FCC to provide a detailed inventory of how public safety spectrum is used and to assess opportunities for reallocation.

H.R. 3630 (§4102) would provide for the eventual release of the narrowband and guard band frequencies for auction to the commercial sector. This auction would be scheduled five years after the Administrator had determined the availability of standards for "public safety voice" (not defined) over the broadband network. Of these auction proceeds, $1 billion would be made available for a grant program for states to purchase radio equipment.

H.R. 3630 would also allow the aggregation of public safety and commercial licenses and the repurposing of commercial spectrum for public safety use.

Management Structure, Network Build-Out and Other Corporate Responsibilities

The administration bills would create a Public Safety Broadband Corporation, which would be charged with the management of the spectrum licenses it would receive, the planning and construction of a national network, and various management and administrative duties, as specified by the bills. Both bills would authorize the establishment of a private, nonprofit corporation established in the District of Columbia. The bills would establish parameters for the membership of the corporation's Board of Directors. Four federal representatives would be designated from the Departments of Commerce, Homeland Security, and Justice, and from the OMB. Eleven nonfederal board members would be appointed by the Secretary of Commerce, in H.R. 2482, and in S. 1549 by the Secretary in consultation with the Secretary of Homeland Security and the U.S. Attorney General. Of the nonfederal board members, at least three would be chosen to represent the collective interests of state, territorial, tribal, and local authorities. In addition to experience in the field of public safety, at least one member of the board would be appointed because of specific abilities in each of these fields: technical expertise, network expertise, and financial expertise. S. 1549 would include experience in cybersecurity as a qualification for technical expertise and would include clauses regarding security clearance that are not mentioned in H.R. 2482.

H.R. 3630 (§4203) would require the NTIA to select, through the process of requests for proposal, an Administrator to provide corporate governance over the deployment and management of interoperable broadband communications for public safety entities as described in the bill. The bill (§4202) would require the FCC to establish a Public Safety Communications Planning Board. Within one year of formation, the board would provide a Public Safety Communications Plan to be carried out by the Administrator. The FCC would provide staff, facilities, and administrative support to the board in its preparation of a plan, as required in the bill. The four federal members would come from the FCC, the NTIA, the OEC, and NIST. Of the remaining nine members of the board, five would represent state and local interests, including three to represent public safety interests, and four members would be from the commercial sector. At least one nonfederal member would be required to provide technical expertise, network expertise, or financial expertise. The FCC would have authority in the actions of the Administrator except to assure that spectrum is used in accordance with the plan, without harmful interference (§4201).

The administration bills would specify the powers and duties of the corporation and would establish statutory responsibilities regarding the build-out, maintenance, and operation of the new public safety network. The bills would permit the corporation to incur debt, including through the sale of notes or bonds.

H.R. 3630 (§4203) would establish the powers of the Administrator, while specifying that those powers are limited to carrying out the requirements of the act and of the Public Safety Communications Plan. The act would establish requirements for state-compliance with the plan regarding the build-out, maintenance, and operation of the new public safety network. The Administrator would be funded solely through the Public Safety Trust Fund.

The administration bills would describe how the Public Safety Broadband Corporation should structure its request for proposals to build out the network. Proposals would be required to include a build-out timetable that included coverage, service levels, and performance criteria. The timetable criteria are similar to what the FCC now requires of commercial network operators through the auction and service rules that are established in advance of any auction of radio frequency spectrum licenses. H.R. 2482 would include additional provisions regarding obligations to seek out commercial partners. S. 1549 would specify construction of "an Evolved Packet Core or Cores," whereas language in H.R. 2482 might be interpreted as specifying the construction of a single core system. The number and placement of operating nodes requiring an Evolved Packet Core is still being debated within the public safety community and its technical advisers.

H.R. 3630 (§4221) would require that states wishing to build a public safety network on the broadband spectrum create a State Public Safety Broadband Office.

The bill would provide requirements to coordinate the responsibilities of the Administrator, the Public Safety Communications Planning Board, and the states.

Under H.R. 3630, each Public Safety Broadband Office would be required to negotiate contracts with a private-sector entity of commercial wireless services. The offices would use a baseline request for proposals prepared by the Public Safety Communications Plan. The negotiated contract would be approved by the Administrator.

The contracts for construction, management, maintenance, and operation of a state public safety broadband communications network are to be in partnership with a commercial provider of mobile services. The contracts would be required to meet the requirements established in the Public Safety Communications Plan.

Public-private partnerships would be permitted and would allow the commercial partner access to spectrum and network resources of the public safety broadband network.

Network access would be permitted on a secondary basis to users that are not part of a public safety entity, as defined in the bill. The Administrator would be required to ensure that any such agreement would allow public safety users pre-emptive access to their network.

Although regional consortia of the State Public Safety Broadband Offices are permitted, H.R. 3630 would not require the construction of, collaboration on, or sharing of evolved packet cores.

The provisions for deployment standards might be interpreted to preclude the Public Safety Communications Plan from including requirements for collaboration or sharing in its plan.

The plan is to require, and the Administrator is to ensure, that each state public safety broadband communications network be interconnected and interoperable with each other.

Provisions for a build-out timetable that would include coverage, service levels, and performance criteria are similar to the administration bills.

Public Safety Roaming and Priority Access

The administration bills would allow the FCC to adopt rules to improve the ability of public safety users to roam onto commercial networks and to gain priority access to commercial networks in an emergency.

H.R. 3630 (§4202) would prohibit any public safety network, device, or other requirement that would be generally applicable to commercial providers of wireless services.

State and Local Implementation

All three bills would establish a State and Local Implementation Fund and would allow the NTIA to borrow against future deposits to the fund from the General Fund of the U.S. Treasury.

The administration bills would provide matching grants to states and localities for some of the costs of connecting to the planned Public Safety Broadband Network.

H.R. 3630 (§4222) would provide matching grants to State Public Safety Broadband Offices to carry out their responsibilities.

Public Safety Research and Development

The administration bills would authorize NIST, in consultation with other federal agencies, to conduct research and to assist with the development of standards, technologies, and applications to advance wireless public safety communications.

H.R. 3630 (§4202) would require states participating in the Public Safety Communications Plan to adhere to the plan's standards for deployment. These would include requirements to use commercial standards and commercial models for network deployment and to be backward-compatible with commercial services.

Funding

The administration bills would require NTIA to manage the initial funding for the start-up costs of the Public Safety Broadband Corporation. S. 1549 would appropriate $50 million for "reasonable expenses." H.R. 2482 would authorize the NTIA to make loans to the corporation; the amount is not specified. In order to make the loans, H.R. 2482 would authorize the NTIA to issues notes or other obligations to the Treasury; the Treasury would be authorized to sell the notes, to be treated as public debt transactions of the United States. Both bills would require the corporation to assess and collect: network user fees; lease fees related to network capacity; lease fees for spectrum used as part of long-haul network transmissions; and lease fees related to network equipment and infrastructure. Both bills would require that the total amount of the fees assessed in each fiscal year shall be sufficient to cover that year's expenses for the corporation.

H.R. 3630 would require the NTIA to advance up to $40 million to the Administrator, repayable from the Public Safety Trust Fund. Total payments from the fund would be $40 million.

The State and Local Implementation Fund would receive $250 million under H.R. 2482; $200 million under S. 1549; and $100 million under H.R. 3630.

Funds from auction proceeds applied to building the Public Safety Broadband Network would total $11.75 billion under H.R. 2482 and $6.45 billion under S. 1549. H.R. 3630 would provide $4.96 billion from initial spectrum auction proceeds for grants to states to build their networks; up to $1.5 billion more might be provided if auction revenues exceed expectations.

Research and development of new technology for public safety would receive $500 million under H.R. 2482 and $300 million under S. 1549. There is no comparable provision in H.R. 3630.

Next Generation 911

Today's 911 system is built on an infrastructure of analog technology that does not support many of the features that most Americans expect to be part of an emergency response. Efforts to splice newer, digital technologies onto this aging infrastructure have created points of failure where a call can be dropped or misdirected, sometimes with tragic consequences. Callers to 911, however, generally assume that the newer technologies they are using to place a call are matched by the same level of technology at the 911 call centers, known as Public Safety Answering Points (PSAPs). To modernize the system to provide the quality of service that approaches the expectations of its users will require investment in new technologies. As envisioned by most stakeholders, these new technologies—collectively referred to as Next Generation 911 or NG9-1-1—should incorporate Internet Protocol (IP) standards.

H.R. 2482 would require two studies intended to expedite the transition to NG9-1-1. The National Highway Traffic Safety Administration would be required to provide Congress with a assessment of NG9-1-1 service readiness, deployment needs, and expected costs. The FCC would be required to prepare a report on legal and statutory changes needed to accommodate the deployment of NG9-1-1.

Table 2. Fund Allocations of Spectrum License Auction Revenue

Fund or Account	S. 1549, Title II, Subt. H	H.R. 2482	H.R. 3630, Title IV
Incentive Auction Relocation Fund	$1 billion	Not less than 5% of Public Safety Trust Fund up to $1 billion	No provisions in bill.
TV Broadcaster Relocation Fund	No provisions in bill.	No provisions in bill.	$3 billion; unexpended funds to be applied to the public safety fund.
Public Safety Trust Fund	All auction proceeds as provided in the bill.	All auction proceeds as provided in the bill.	All auction proceeds as provided in the bill except as noted for TV.
State and Local Implementation	$200 million	$250 million	$100 million
Public Safety Broadband Corporation	$50 million in appropriations.	Loan against future revenue.	No provisions in bill.
Public Safety Administrator	No provisions in bill.	No provisions in bill.	$40 million from the public safety fund.
Public Safety Broadband Network Deployment	$6.45 billion	$11.75 billion	$4.96 billion plus 10% of any remaining amounts deposited in the fund up to $1.5 bn.
Payments from Public Safety Trust Fund for Deficit Reduction	Balances remaining when fund expires.	Balances remaining when fund expires.	$20.4 billion from fund and balances upon expiration plus at least 90% of any additional auction revenue.
Public Safety Research	$300 million	$500 million	No provisions in bill.
Other research	No provisions in bill.	$350 million	No provisions in bill.
9-1-1 grant program	No provisions in bill.	No provisions in bill.	$250 million
Public Safety Trust Fund Expires	FY2018	FY2021	FY2021

H.R. 3630 (§4202) would require that states building out broadband networks in accordance with the act connect to 911 call centers and NG9-1-1 networks. The bill would incorporate the provisions of the Next Generation 9-

1-1 Preservation Act of 2011 (H.R. 2629, Representative Shimkus). These provisions would re-authorize the 9-1-1 Implementation Coordination Office and its programs as established by P.L. 108-494 and P.L. 110-283. These provisions would include planning for NG9-1-1 and a grant program that would receive $250 million from the Public Safety Trust Fund.

APPENDIX A. FURTHER DISCUSSION OF LEGISLATION IN THE 112TH CONGRESS TO IMPROVE EMERGENCY COMMUNICATIONS

A number of proposed bills in the 112th Congress commonly depend on federal grant guidance and rule-making to bring about needed changes. The bills would help to move public safety agencies away from the incremental development of narrowband voice networks to next-generation IP-enabled networks that fully support voice, data, and video transmissions.

They all would assign the D Block license to the Public Safety Broadband Licensee for a new broadband network. S. 28, S. 1040, and H.R. 607 would give the FCC a central role in decision-making and planning for network construction. S. 1040 and H.R. 607 would include DHS in planning efforts and assign to it the responsibility for program grants. S. 911, as amended, is one of several bills that would give grant-making authority to the NTIA.

Funding for the grants would come from auctioning spectrum licenses; loans to grants programs could be substantial, based on expected auction proceeds. S. 911, as amended, H.R. 2482, S. 1549, and H.R. 12 would create a new not-for-profit corporation that would plan for and develop a nationwide network.

These bills include the Public Safety Spectrum and Wireless Innovation Act (S. 28, Rockefeller), the Broadband for Public Safety Act of 2011 (S. 1040, Lieberman), the Strengthening Public-safety and Enhancing Communications Through Reform, Utilization, and Modernization (SPECTRUM) Act (S. 911, Rockefeller, as amended), the Public Safety and Wireless Innovation Act (H.R. 2482, Dingell), the Broadband for First Responders Act (H.R. 607, King), and the American Jobs Act of 2011, Title II, Subtitle H (S. 1549, Reid and H.R. 12, Larson).

Public Safety Spectrum and Wireless Innovation Act, S. 28

Under this proposal, governance of a public safety broadband network would be shared by several agencies. S. 28 would require the formation of an advisory board with which the FCC would consult.[94] Annual appropriations through 2018 would be authorized for the Emergency Response Interoperability Center to carry out its responsibilities as established in the bill.[95] The NTIA would be responsible for administering the grants program for network construction, in consultation with the FCC, which would define project requirements.[96] The FCC would administer a separate maintenance and operation reimbursement fund. It would require annual status reports from license-holders.[97] The Comptroller General would be required to perform audits of the construction fund and the maintenance and operation fund.[98]

S. 28 would empower the FCC to take "all actions necessary" to ensure the deployment of the public safety broadband network.[99] It would determine whether spectrum licenses would be national, regional, or statewide, which would influence sharing of decision-making powers,[100] and would authorize partnerships with commercial interest to build a state's public safety broadband network.[101] It would set standards for, authorize, and to some extent supervise requests for proposals to build networks.[102] The bill would establish specific requirements concerning network build-out in rural areas,[103] and requirements for assistance from the General Services Administration.[104]

Sources of revenue would come from the proceeds of commercial spectrum auctions, as described in the bill,[105] and from fees or other income from secondary users of public safety spectrum.[106] Auction proceeds up to $11 billion would be divided between a Construction Fund and a Maintenance and Operation Fund, established by the act.[107] Proceeds above $11 billion would go to "growth-enhancing" infrastructure projects.[108] An antidiversion prohibition would require that any funds for public safety programs made available through provisions of the act would be spent in accordance with FCC guidelines.[109]

S. 28 would direct the FCC to reallocate the D Block to public safety use[110] and to determine the assignment of licenses.[111] It would set up rules to govern the authorization of secondary access, if any, for those licenses.[112] The FCC would also establish rules for public safety to access commercial spectrum and infrastructure, including roaming and priority access.[113] It would determine whether the public safety licenses now designated for narrowband network use might be used for broadband technologies and set the requirements for this usage.[114] Within five years of the enactment of the act, the FCC would be required to report to Congress on how public safety agencies are using public safety spectrum allocations and whether more spectrum should be made available.[115]

The FCC would set technical and operational rules for the network[116] and, with NIST, develop standards necessary to ensure interoperability, security, and functionality.[117] The bill would require a GAO report on incorporating satellite communications into the broadband network.[118]

To capture spectrum license auction revenues, S. 28 would extend the auction authority from the end of FY2012 to FY2020. In addition to specifically requiring an auction of at least 25 MHz of spectrum from frequencies between 1675 and 1710 MHz and spectrum between 2155 and 2180 MHz, the bill would give authority to the FCC to organize and conduct voluntary auctions.[119] The FCC has specifically requested this authority primarily so that it may provide financial incentives to television broadcasters as part of a plan to repurpose some of the broadcaster spectrum holdings.

Spectrum Act, S. 911 as amended, and H.R. 2482

As amended by a substitute amendment and other amendments, S. 911 and its companion bill, H.R. 2482, would create a private, non-profit Public Safety Broadband Corporation to provide governance for a nationwide interoperable network for public safety communications.[120] The board of directors of the corporation would include the Secretaries of Commerce and Homeland Security, the Attorney General, and the Director of the Office of Management and Budget, or their designees. Eleven nonfederal members would also sit on the board.[121]

The corporation would be tasked with deploying and operating the nationwide network according to provisions established in the bills.

The bills would create a Public Safety Trust Fund that would receive proceeds of auctions of commercial spectrum licenses, including federal spectrum repurposed for commercial use, as described in the bill.[122] The Public Safety Broadband Corporation would receive grants through the fund. The bills would require a grant to the corporation of $11.75 billion, of which at least $10.5 billion would be available for the radio network and at least $1.25 billion would be available to develop the standards and technology for the network's evolved packet core.[123] A State and Local Implementation Fund of $250 million would be created to distribute grants from the fund for the purpose of connecting to the nationwide network.[124] The Corporation would be required to develop its own revenue streams through assessing and collecting fees for network use by third parties.[125]

The bills would direct the FCC to reallocate the D Block to public safety use and to assign it and the existing public safety broadband license to the newly created corporation.[126] The bills would also authorize the FCC to allow narrowband spectrum to be used flexibly, including for public safety broadband communications.[127]

The Public Safety Trust Fund would be required to provide grants to NIST for research and development for various purposes as required by the bills, including for public safety needs.[128] The fund would also be used for wireless communications research to be undertaken by the National Science Foundations and the Defense Advanced Research Projects Agency (DARPA).[129]

To capture spectrum license auction revenues, the bills would extend the auction authority from the end of FY2012 to FY2021.[130] The bills would also authorize the FCC to conduct incentive auctions.[131] At least 5%, but no more than $1 billion of auction proceeds, would be available for an Incentive Auction Relocation Fund. These funds would be available to facilitate the voluntary relinquishment of spectrum licenses. The bills also identify radio frequencies that are to be made available for auction and would require the NTIA to make available at least 15 MHz of contiguous government-occupied spectrum between the 1675 to 1710 MHz band within one year of enactment of the bill.[132]

American Jobs Act of 2011, Subtitle H (S. 1549 and H.R. 12)

S. 1549 and its companion bill, H.R. 12, would create a private, non-profit Public Safety Broadband Corporation to provide governance for a nationwide interoperable network for public safety communications.[133] The board of directors of the corporation would include the Secretaries of Commerce and Homeland Security, the Attorney General, and the Director of the Office of Management and Budget, or their designees. Eleven nonfederal members would also sit on the board.[134]

The corporation would be tasked with deploying and operating the nationwide network according to provisions established in the bills.[135]

The bills would create a Public Safety Trust Fund that would receive proceeds of auctions of commercial spectrum licenses, including federal spectrum repurposed for commercial use, as described in the bill.[136] The

Public Safety Broadband Corporation would receive grants through the fund.[137] The bills would require funding for the corporation of $6.45 billion, after submission of a five-year budget to be approved by the Secretary of Commerce, in consultation with the Secretary of Homeland Security, the Director of the Office of Management and Budget, and the Attorney General of the United States.[138] A State and Local Implementation Fund of $200 million would be created to distribute grants from the fund for the purpose of connecting to the nationwide network.[139] The Corporation would be required to develop its own revenue streams through assessing and collecting fees for network use by third parties.[140]

The bills would direct the FCC to reallocate the D Block for use by public safety entities.[141] The bills would also authorize the FCC to allow narrowband spectrum to be used flexibly, including for public safety broadband communications.[142]

Up to $300 million would be made available for a Wireless Innovation Fund that might go to NIST for public safety research and development. The Director of the Office of Management and Budget would be required to approve the spend plan. If less that $300 million is provided to NIST, the remainder would go to the Public Safety Broadband Corporation.[143]

To capture spectrum license auction revenues, the bills would make the FCC's auction authority permanent.[144] The bills would also authorize the FCC to conduct incentive auctions if consistent with the public interest.[145] No more than $1 billion of auction proceeds would be available for an Incentive Auction Relocation Fund for incentive auctions.[146] The bill would require the NTIA to make available at least 15 MHz of contiguous government-occupied spectrum between the 1675 to 1710 MHz band within one year of enactment of the bill.[147] Additional frequencies are also identified for possible auction.[148]

Broadband for Public Safety Act of 2011, S. 1040

Under S. 1040, governance would come primarily from grants administration. DHS would administer grants programs set up through the creation of a Construction Fund.[149] DHS would also administer a Maintenance and Operation Fund.[150] The Comptroller General would be required to perform audits.[151] The FCC would retain its power to establish the roles and responsibilities of the PSBL.[152] The bill would expand the Board of Directors of the PSBL, mandating the appointment of representatives from over 40 associations listed in the bill and allowing for the election of additional members.[153]

S. 1040 would require each state to provide information about its plans for deployment of the new network;[154] the plans would be submitted to DHS and the FCC for joint review and approval.[155] Grants could be applied to improvements and new infrastructure to meet public safety requirements for networks in the 700 MHz band, the 800 MHz band, or the 4.9 GHz band.[156] The OEC would advise construction grant recipients on best practices and would provide guidance of project implementation.[157]

The Construction Fund and the Maintenance and Operation Fund would receive proceeds from spectrum license auctions, as designated in the bill.[158] The first $5.5 billion from designated auctions would be deposited in the Construction Fund.[159] Auction proceeds in excess of $5.5 billion, up to $11 billion, would go to the Maintenance and Operation Fund.[160] Auction revenues above $11 billion would be applied to deficit reduction.[161] Any unspent money in the Construction Fund would be transferred to the Maintenance and Operation Fund at the conclusion of the construction phase; the end of the construction phase would be determined by DHS.[162] Any revenue from sharing, leasing, or sublicensing access to the public safety spectrum licenses or infrastructure would be deposited in the Maintenance and Operation Fund.[163]

The FCC would be directed to reallocate the D Block for public safety use and assign the license to the PSBL.[164] The PSBL would be permitted to authorize providers of public safety services to construct and operate wireless broadband public safety networks on its spectrum holdings.[165] The FCC would be required to authorize shared use, sublicensing, or leasing, provided that public safety services receive priority access to the network.[166] Spectrum allocated for public safety use at 4.9 GHz would be opened to restricted sharing with commercial users.[167] The FCC, in consultation with NIST, DHS, and others, would also set rules for interoperability between public safety and commercial networks and for roaming.[168] Upon enactment of the bill, the FCC would be required to end the renewal of public safety licenses between 170 MHz and 512 MHz except under circumstances specified in the bill.[169] Within three years, the GAO would be required to submit a report to Congress identifying public safety spectrum holdings that could be reassigned through auction and the likely cost of such a migration.[170] The FCC would be required to submit a report to Congress on how public safety agencies are using public safety spectrum allocations and whether more spectrum should be made available.[171] Within five years, the FCC and other stakeholders would be required to provide a recommendation to Congress regarding the transfer of communications systems below 512 MHz to public safety licenses in the 700 MHz and 800 MHz bands.[172] Funding for this migration might be made available from the Maintenance and Operation Fund.[173] The bill would mandate that all federal law enforcement agency communications not operating on commercial networks transfer to frequencies in the 700 MHz and 800 MHz bands, within 10 years of the date of enactment of the bill.[174]

The bill specifies minimum requirements to be established by the FCC for the broadband network.[175] DHS, in consultation with NIST, shall establish standards to meet the public safety requirements developed by the FCC.[176] The FCC, in consultation with DHS and the NTIA, would be required to issue a report and order on the use of IP-enabled networks to achieve interoperability.[177]

To capture spectrum license auction revenues, S. 1040 would extend the auction authority from the end of FY2012 to FY2020.[178] An auction of at least 15 MHz of contiguous spectrum from frequencies between 1675 and 1710 MHz would be required within a year of enactment of the bill.[179] Additional frequencies are identified for auction not later than January 31, 2014.[180]

Broadband for First Responders Act, H.R. 607

H.R. 607 would expand the Board of Directors of the PSBL, mandating the appointment of representatives from 40 associations listed in the bill and allowing for the election of additional members.[181] The licensee would be required to submit a report to Congress on network plans.[182] The Comptroller General would be required to perform audits.[183]

The FCC would establish the rules for public safety service providers to construct and operate a network on determination by the PSBL that this action would expedite network deployment.[184] DHS would administer grants programs set up through the creation of a Construction Fund and a Maintenance and Operation Fund.[185]

The act would create a Construction Fund and a Maintenance and Operation Fund to receive proceeds from spectrum license auctions.[186] The first $5.5 billion from designated auctions would be deposited in the Construction Fund, with the balance going to the Maintenance and Operation Fund.[187] Appropriations of up to $5.5 billion would be authorized to supplement auction revenue, if needed to reach a total of $11 billion.[188]

H.R. 607 would direct the FCC to reallocate the D Block to public safety use and permit access to the public safety broadband spectrum and infrastructure to other providers.[189] The bill would require a public safety agency statement of requirements that would enable nationwide interoperability and roaming across any communications system that used public safety spectrum, as defined in the law.[190] The bill would mandate the transition from narrowband systems below 512 MHz to networks operating on public safety frequencies in the 700 MHz and 800 MHz bands.[191] The FCC and other stakeholders would be required to provide a detailed plan for the transition.[192]

The FCC would set technical and operational rules for the network[193] and, with NIST and others, set requirements to ensure interoperability, security, and functionality.[194] DHS, with NIST, would take the lead in developing standards to meet these and other requirements.[195] The FCC, in consultation with DHS and the NTIA, would be required to issue a report and order on the use of IP-enabled networks to assist interoperability.[196]

To capture spectrum license auction revenues, H.R. 607 would extend the auction authority from the end of FY2012 to FY2020.[197] The GAO would be required to submit a report identifying the frequencies below 512 MHz used by public safety agencies that should be reassigned for auction as commercial licenses.[198] The bill would require the recovery of spectrum at 420-440 MHz and 450-470 MHz for auction and encourage the FCC to reconfigure spectrum to increase the value of these bands.[199] Some of this spectrum is currently assigned for amateur radio use.

To capture spectrum license auction revenues, the draft discussion bill might extend the authority of the FCC from the end of FY2012 to FY2021. Frequencies used by the federal government are identified for possible auction. The FCC might be given the authority to establish incentive auctions; statutory protections regarding TV broadcasters participating in or affected by incentive auctions are proposed. Other provisions regarding auctions might be applied.

APPENDIX B. PROPOSALS FOR SPECTRUM ASSIGNMENT

The Federal Communications Commission (FCC)[200] is the only federal agency to propose a national network for public safety communications infrastructure and to take action to plan, implement, and fund it. The FCC has proposed a public-private partnership to build a broadband network to benefit public safety.[201] Its proposal would incorporate obligations into auction rules for a commercial network operator that would provide for a shared, national network using public safety and commercial frequencies at 700 MHz. Planning for the network would be conducted on a nationwide basis. To create the partnership, the FCC provided for two national licenses of 10 MHz each. One license was assigned to a Public Safety Broadband Licensee (PSBL). The 10- MHz license held by the PSBL is part of the 24 MHz originally assigned to public safety. The other license—designated the Upper Block D, or D Block—was scheduled for auction in 2008 to a commercial provider. At the auction, the FCC's requirements could not be met by any bidder. The failure to find a commercial partner to work with the public safety license-holder effectively reset the planning process to zero. The FCC decided to include an assessment of public safety broadband needs as part of its preparation of a national broadband plan, as required by the American Recovery and Reinvestment Act.[202] With a public notice released September 28, 2009,[203] the FCC sought information about current and potential future use of broadband in public safety communications.

As stated above, after the auction of the D Block failed in early 2008, the FCC issued a new request for comments on how to restructure the auction to provide a network that would meet public safety needs. This led to a host of new suggestions on how to use the spectrum. The FCC's choices, however, are constrained by provisions of the Deficit Reduction Act of 2005 that require it to auction the D Block. Many of the options proposed to the FCC might therefore require Congress to amend the act or to introduce other enabling legislation.

Comments filed with the FCC[204] have opened debates about alternative courses of action. Although there are a number of different proposals, each in some way addresses the question of whether it will be public safety representatives or commercial owners and network operators that control the decision-making process. The following is a summary of proposals under discussion and possible agency or legislative actions that might be needed to implement them.[205] Other options may be proposed or developed.

Auction the D Block to Commercial Interests

- Agree to new rules for a D Block auction that satisfy the key goals laid out by the FCC for a shared network that benefits both public safety users and commercial interests. The D Block might be auctioned as a single, national license or as many licenses assigned to specific geographic areas. The FCC has the authority to structure such an auction under the Communications Act of 1934, as amended, including the amendments provided by the Deficit Reduction Act of 2005. This approach was recommended by the FCC.

- Auction the D Block without any obligations to share with public safety. Public safety agencies could eventually add broadband applications to communications systems built in the 24 MHz of frequencies originally assigned to them. The FCC has the authority to facilitate this decision.
- Auction the D Block without any obligations to share with public safety **and "give" the auction proceeds to public safety**, possibly by transferring it to an existing grants program. One program that has been mentioned is the Public Safety Interoperable Communication (PSIC) grant program set up by a provision in the Deficit Reduction Act of 2005. Congress might choose to amend the provisions for grants in the Deficit Reduction Act of 2005 or it might choose to create new legislation specifically for the distribution of the D Block auction proceeds. In either case, the cost of new authorizations would be scored by the Congressional Budget Office. Except where Congress has provided for exceptions, the Communications Act of 1934 states that, as a general rule, auction proceeds are deposited in the U.S. Treasury.[206]

Assign the D Block to Public Safety Licensees

- Assign the D Block to the Public Safety Broadband Licensee to administer as part of plans for a nationwide, interoperable broadband network for public safety. Because the Deficit Reduction Act of 2005 included the D Block frequencies among those that the FCC was required to auction, it appears that Congress must amend the law in order for this particular option to be exercised.
- Assign the D Block to the Public Safety Broadband Licensee for administration and provide federal grants to state, local or regional agencies to build and operate the network. This approach, which would include spectrum license auctions as a source of revenue, has been endorsed by the Obama administration and is included in its 2012 Budget.[207]
- Assign the D Block to state, local, or regional network managers. These would negotiate with commercial partners on how to build,

operate, and fund a shared network. This action would require agreement among the participants on how to assure nationwide interoperability of the separate networks. The Public Safety Broadband Licensee would provide matching sub-licenses for its spectrum holdings in those areas.

- Combine the D Block with the existing public safety broadband license and award the license to a new not-for-profit corporation set up for the purpose of building and administering the public safety broadband network. Funding sources for the network would be a mix of federal and private sector investment.

Auction the D Block and the Public Safety Broadband Network Licenses

Combine the 10-MHz D Block with the 10-MHz Public Safety Broadband License and auction the newly created 20-MHz block, with sharing requirements similar to some of those considered by the FCC for the D Block auction. Funds from the auction proceeds might go to public safety through a program established for that purpose. In addition to legislation that would establish the funding program, Congress might need to amend language in the Balanced Budget Act of 1997 that directed the FCC to assign 24 MHz from the 700 MHz band to public safety.

APPENDIX C. CONGRESSIONAL EFFORTS ON BEHALF OFPUBLIC SAFETY COMMUNICATIONS

Many of the statutes passed since 2001 have provided guidelines and set performance goals for public safety communications while delegating decisions about implementation to federal agencies and state officials. Although Congress has appropriated money for public safety communications, it has not directly addressed the question of investment in network infrastructure, leaving it largely to federal agencies to set priorities for how public safety grants can be used. Grants for emergency communications have been used to purchase equipment that facilitates interoperability, for planning, and for training.

Congress first addressed the issue of emergency communications interoperability in the Balanced Budget Act of 1997 by providing additional radio frequency spectrum that would allow for interoperable networks. Provisions intended to improve interoperable functions in public safety networks were included in the Homeland Security Act of 2002 (P.L. 107-296). Two years later, responding to recommendations of the National Commission on Terrorist Attacks Upon the United States (9/11 Commission), Congress included a section in the Intelligence Reform and Terrorism Prevention Act of 2004 (P.L. 108-458) that expanded its requirements for action in improving interoperability and public safety communications. Also in response to a recommendation by the 9/11 Commission, Congress set a firm deadline for the release of radio frequency spectrum needed for public safety radios, as part of the Deficit Reduction Act of 2005 (P.L. 109-171). These laws provided the base from which the Department of Homeland Security (DHS) might develop a national public safety communications capability as required by the Homeland Security Appropriations Act, 2007 (P.L. 109-295). Title VI, Subtitle D of the act, referred to as the 21st Century Emergency Communications Act of 2006, placed new requirements on DHS. Additional requirements were included in the Implementing Recommendations of the 9/11 Commission Act of 2007 (P.L. 110-53).

Balanced Budget Act of 1997

The initial allocation to public safety of frequencies in the 700 MHz band was required by Congress in the Balanced Budget Act of 1997 (P.L. 105-33),[208] which directed the Federal Communications Commission (FCC) to designate 24 MHz of spectrum capacity for public safety. To carry out the process of assigning this newly allocated spectrum asset, the FCC created the Public Safety National Coordination Committee (NCC) as a Federal Advisory Committee. Active from 1999 through 2003, the NCC had a Steering Committee from government, the public safety community, and the telecommunications industry. The NCC developed technical and operational recommendations for the 700 MHz band, including plans for interoperable channels. The existing governance for these channels is through Regional Planning Committees (RPCs),[209] established and loosely coordinated by the FCC, with the participation of the National Public Safety Telecommunications Council (NPSTC), a group consisting primarily of public safety associations.

The RPCs are responsible for submitting 700 MHz band plans to the FCC for approval, and for managing these plans.

The Homeland Security Act of 2002 and Actions by the Department

Provisions of the Homeland Security Act instructed DHS to address some of the issues concerning public safety communications in emergency preparedness and response and in providing critical infrastructure. Telecommunications for first responders is mentioned in several sections, with specific emphasis on technology for interoperability.[210]

The newly created DHS placed responsibility for interoperable communications within the Directorate for Science and Technology, reasoning that the focus of DHS efforts would be on standards and on encouraging research and development for communications technology. Responsibility to coordinate and rationalize federal networks, and to support interoperability, had previously been assigned to the Wireless Public SAFEty Interoperable Communications Program—called Project SAFECOM—by the Office of Management and Budget as an e-government initiative. With the support of the George W. Bush administration, SAFECOM was placed in the Science and Technology directorate and became the lead agency for coordinating federal programs for interoperability.[211] The Secretary of Homeland Security assigned the responsibility of preparing a national strategy for communications interoperability to the Office of Interoperability and Compatibility (OIC), which DHS created, an organizational move that was later ratified by Congress in the Intelligence Reform and Terrorism Prevention Act.[212] SAFECOM continued to operate as an entity within the OIC, which assumed the leadership role.

Intelligence Reform and Terrorism Prevention Act

Acting on recommendations made in 2004 by the 9/11 Commission, Congress included several sections regarding improvements in communications capacity—including clarifications to the Homeland Security Act—in the Intelligence Reform and Terrorism Prevention Act (P.L. 108- 458).

The Commission's analysis of communications difficulties on September 11, 2001, was summarized in the following recommendation.

> Congress should support pending legislation which provides for the expedited and increased assignment of radio spectrum for public safety purposes. Furthermore, high-risk urban areas such as New York City and Washington, D.C., should establish signal corps units to ensure communications connectivity between and among civilian authorities, local first responders, and the National Guard. Federal funding of such units should be given high priority byCongress.[213]

Congress addressed both the context and the specifics of the recommendation for signal corps capabilities. The Intelligence Reform and Terrorism Prevention Act amended the Homeland Security Act to specify that DHS give priority to the rapid establishment of interoperable capacity in urban and other areas determined to be at high risk from terrorist attack. The Secretary of Homeland Security was required to work with the Federal Communications Commission (FCC), the Secretary of Defense, and the appropriate state and local authorities to provide technical guidance, training, and other assistance as appropriate. Minimum capabilities were to be established for "all levels of government agencies," first responders, and others, including the ability to communicate with each other.[214] The act further required the Secretary of Homeland Security to establish at least two trial programs in high-threat areas. The process of development for these programs was to contribute to the creation and implementation of a national model strategic plan. The purpose was to foster interagency communications at all levels of the response effort. Building on the concept of using the Army Signal Corps as a model, the law directed the Secretary to consult with the Secretary of Defense in the development of the test projects, including review of standards, equipment, and protocols.[215]

Congress also raised the bar for performance and accountability, setting program goals for the Department of Homeland Security. Briefly, the goals were to:

- Establish a comprehensive, national approach for achieving interoperability;
- Coordinate with other federal agencies;
- Develop appropriate minimum capabilities for interoperability;
- Accelerate development of voluntary standards;
- Encourage open architecture and commercial products;
- Assist other agencies with research and development;

- Prioritize, within DHS, research, development, testing and related programs;
- Establish coordinated guidance for federal grant programs;
- Provide technical assistance; and
- Develop and disseminate best practices.

The act included a requirement that any request for funding from DHS for interoperable communications "for emergency response providers" be accompanied by an Interoperable Communications Plan, approved by the Secretary. Criteria for the plan were also provided in the act.[216]

The act also provided a sense of Congress that the next Congress—the 109th—should pass legislation supporting the Commission's recommendation to expedite the release of spectrum. This was addressed in the Deficit Reduction Act of 2005 (P.L. 109-171).

Deficit Reduction Act of 2005 and the Public Safety Interoperability Grant Program

Provisions in the Deficit Reduction Act of 2005 planned for the release of spectrum by February 18, 2009,[217] and created a fund to receive spectrum auction proceeds and disburse designated sums to the Treasury and for other purposes,[218] including a grant program of up to $1 billion for public safety agencies. The fund's disbursements were to be administered by the NTIA.[219] At the time, the Congressional Budget Office projected that the grants program for public safety would receive $100 million in FY2007, $370 million in FY2008, $310 million in FY2009 and $220 million in FY2010.[220] However, the 109th Congress, in its closing hours, passed a bill with a provision requiring that the grants program receive "no less than" $1 billion to be awarded "no later than" September 30, 2007.[221] Language in Implementing Recommendations of the 9/11 Commission Act of 2007 (P.L. 110-53) required some changes in the grant program and reaffirmed the 2007 fiscal year deadline.[222]

In February 2007, the NTIA transferred the management of the public safety grant program to DHS, signing a memorandum of understanding (MOU) with the Office of Grants and Training.[223] The MOU included an overview of how the Public Safety Interoperable Communications (PSIC) Grant Program, as it is called, is to be administered. The overview was reiterated and explained in testimony.[224] Both the MOU and the testimony indicate that the priority was to fund needs identified through Tactical Interoperable Communications Plans and Statewide Interoperable Plans developed in conjunction with SAFECOM.

On July 18, 2007, the Secretaries of Commerce and Homeland Security jointly announced the details of the PSIC grant program.[225] The program, as announced, was to provide $968,385,000 in funding for all 50 states, the District of Columbia, and U.S. Territories.[226] The announcement of the top-level, statewide allocations met the September 30 deadline set by Congress. The states were required to submit brief descriptions of envisioned projects and how grant requirements and guidelines would be met. One of the requirements was that states must have a Statewide Communication Interoperability Plan (SCIP) that has been approved by DHS's Office of Emergency Communications. Actual expenditure amounts are reported as states tap their allocations. Under current law, the states and eligible territories have until the end of FY2011, with a possible extension to 2012, to use the funds made available to them.[227]

In 2010, an audit by the Office of the Inspector General (OIG), Department of Commerce, provided PSIC grant amounts and drawdowns by state through March 31, 2010. At that time the drawdowns amounted to 31% of the $968.4 million made available through the PSIC program.[228] The OIG has undertaken audits of PSIC grants in nine states that provide some information on how the grant money is being used.[229]

The Homeland Security Appropriations Act, 2007

The destruction caused by Hurricanes Katrina and Rita in August-September 2005 reinforced the recognition of the need for providing interoperable, interchangeable communications systems for public safety and also revealed the potential weaknesses in existing systems to withstand or recover from catastrophic events. Testimony at numerous hearings following the hurricanes suggested that DHS was responding minimally to congressional mandates for action, most notably as expressed in the language of the Intelligence Reform and Terrorism Prevention Act. Bills subsequently introduced in both the House and the Senate proposed strengthening emergency communications leadership and expanding the scope of the efforts for improvement. Some of these proposals were included in Title VI of the Homeland Security Appropriations Act, 2007 (P.L. 109-295). Title VI—the Post-Katrina Emergency Management Reform Act of 2006 reorganized the Federal

Emergency Management Agency (FEMA), gave the agency new powers, and clarified its functions and authorities within DHS.[230]

The act also addressed public safety communications in Title VI, Subtitle D—the 21st Century Emergency Communications Act of 2006. This section created an Office of Emergency Communications (OEC) and the position of Director, reporting to the Assistant Secretary for Cybersecurity and Communications. The Director was required to take numerous steps to coordinate emergency communications planning, preparedness, and response, particularly at the state and regional level. These efforts were to include coordination with Regional Administrators appointed by the FEMA Administrator to head 10 Regional Offices. To assist these efforts, Congress required the creation of Regional Emergency Communications Coordination (RECC) Working Groups.[231]

Other responsibilities assigned to the Director included conducting outreach programs, providing technical assistance, coordinating regional working groups, promoting the development of standard operating procedures and best practices, establishing nonproprietary standards for interoperability, developing a national communications plan, working to assure operability and interoperability of communications systems for emergency response, and reviewing grants. Required elements of the National Emergency Communications Plan (NECP) included establishing requirements for assessments and reports, and an evaluation of the feasibility of developing a mobile communications capability modeled on the Army Signal Corps. The feasibility study was to be done by DHS on its own or in cooperation with the Department of Defense. Congress also required assessments of emergency communications capabilities, including an inventory that identified radio frequencies used by federal departments and agencies.[232] The completed National Emergency Communications Plan set goals for improving emergency communications and interoperability but did not address developing a network infrastructure for public safety communications or for using the 700 MHz spectrum for that purpose.[233]

To support its vision of interoperability as a system of systems, DHS sponsored an Emergency Response Council (ERC) composed of several dozen agencies, associations, and other entities involved in public safety and emergency response planning.[234] In 2007 the ERC provided a set of agreements on a Nationwide Plan for Interoperable Communications. The ERC published 12 guiding principles deemed essential to their key goals of forging partnerships, designing interoperable systems, educating policymakers, and allocating resources.[235] To date, the council's role has been primarily to establish a base for advocacy and communication among representatives of public safety agencies and associations.

Regional Emergency Communication Coordination

In P.L. 109-295, Congress directed the OEC to coordinate with the Regional Emergency Communication Coordination (RECC) Working Groups established by FEMA.[236] These groups were to provide a platform for coordinating emergency communications plans among states and were intended to include representatives from many sectors with responsibility for public safety and security. Plans for forming RECCs were announced in December 2007. In 2008 organization charts were developed, graphing how the RECCs were structured and where they would fit in the existing chain-of-command of the Federal Emergency Management Agency (FEMA). A National RECC Coordinator was appointed and plans were announced to appoint administrators for each of the regions.

A key proposal for the RECC structure is to "Establish and use the RECC's as a single Federal emergency communications coordination point for Federal interaction with the State, local and tribal governments."[237] Congress placed an emphasis on assisting first responders in its statement of RECC goals but did not limit the RECCs' ability to set more inclusive goals. Although the RECCs might be an effective conduit for interaction to develop policies and plan for shared infrastructure, they are currently used primarily as a forum for FEMA's Disaster Operations Directorate to relay guidelines and orders. Based on the role of RECCs as assigned by the National Emergency Communications Plan, their focus will be on assisting first responders to prepare for disaster response. Leadership will be provided by FEMA and governance will be through the chain-of-command of the agencies' directorates.

The formation of the regional working groups, the RECCs, responded in part to requests from the public safety community to expand interoperable communications planning to include the second tier of emergency workers. Nonfederal members of the RECC are to include first responders, state and local officials and emergency managers, and public safety answering points (911 call centers). Additionally, RECC working groups are to coordinate with a variety of communications providers (such as wireless carriers and cable operators), hospitals, utilities, emergency evacuation transit services, ambulance services, amateur radio operators, and others as appropriate.

National Emergency Communications Plan

In compliance with requirements of the Homeland Security Appropriations Act, 2007, the Department of Homeland Security issued the National Emergency Communications Plan (NECP) in July 2008.[238]

The NECP sets three goals for levels of interoperability[239]

- By 2010, 90% of all areas designated within the Urban Areas Security Initiative (UASI) will demonstrate response-level emergency communications, as defined in grant programs, within one hour for routine events involving multiple jurisdictions and agencies.
- By 2011, 75% of non-UASI will have achieved the goal set for UASIs.
- By 2013, 75% of all jurisdictions will be able to demonstrate response-level emergency communications within three hours for a significant incident as outlined in national planning scenarios.

These jurisdictional goals are to be knit together into a national communications capability through program efforts such as FEMA's Regional Emergency Communications Coordination (RECC) Working Group. The three goals are bolstered by seven objectives for improving emergency communications for first responders, dealing largely with organization and coordination.[240] Each of these objectives has "Supporting Initiatives" and milestones.

End Notes

[1] Some of the actions by Congress and by Federal agencies have been discussed previously in these and other CRS reports: CRS Report R40859, *Public Safety Communications and Spectrum Resources: Policy Issues for Congress* , by Linda K. Moore; CRS Report RL34054, *Public-Private Partnership for a Public Safety Network: Governance and Policy,* by Linda K. Moore; CRS Report RL33838, *Emergency Communications: Policy Options at a Crossroads,* by Linda K. Moore; and CRS Report RL32594, *Public Safety Communications Policy,* by Linda K. Moore.

[2] A frequently cited definition of interoperability has been provided by SAFECOM, an agency with the Department of Homeland Security: "In general, interoperability refers to the ability of public safety emergency responders to work seamlessly with other systems or products without any special effort. Wireless communications interoperability specifically refers to the ability of public safety officials to share information via voice and data signals on

demand, in real time, when needed, and as authorized." See http://www.safecomprogram.gov/SAFECOM/interoperability/default.htm.

[3] Hearings in the 112th Congress include Senate, Committee on Science, Commerce and Transportation, "Safeguarding our Future: Building a Nationwide Network for First Responders," February 16, 2011; House, Committee on Homeland Security, "Public Safety Communications: Are the Needs of Our First Responders Being Met?," March 30, 2011, House, Committee on Energy and Commerce, Subcommittee on Communications and Technology, "Using Spectrum to Advance Public Safety, Promote Broadband, Create Jobs, and Reduce the Deficit," April 12, 2011; House, Committee on Energy and Commerce, Subcommittee on Communications and Technology, "Creating an Interoperable Public Safety Network," May 24, 2011, House, Committee on Energy and Commerce, Subcommittee on Communications and Technology, "Legislative Hearing to Address Spectrum and Public Safety Issues," July 15, 2011, House, Committee on Homeland Security, "Hearing on the Attacks of September 11: Where are We Today," September 8, 2011, House, Committee on Energy and Commerce, Subcommittee on Communications and Technology, markup of the Discussion Draft of the Jumpstarting Opportunity with Broadband Spectrum (JOBS) Act of 2011, December 1, 2011.

[4] The White House, Office of the Press Secretary, "President Obama Details Plan to Win the Future through Expanded Wireless Access" February 10, 2011, http://www. whitehouse.gov/the-press-office/2011/02/10/president-obama-detailsplan-win-future-through-expanded-wireless-access.

[5] Public Safety Alliance, at http://www.psafirst.org/.

[6] FCC, *Connecting America: The National Broadband Plan:* http://www.broadband. gov/download-plan/.

[7] The Connect Public Safety Now coalition is one example; see http://www. Connectpubli csafetynow.org/issue.

[8] House, Committee on Energy and Commerce, Subcommittee on Communications and Technology, markup of the Discussion Draft of the Jumpstarting Opportunity with Broadband Spectrum (JOBS) Act of 2011, December 1, 2011.

[9] According to *Interoperable Communications for First Responders,* an undated white paper from Frost & Sullivan, interoperable communications for first responders was first addressed in 1970, when the FIRESCOPE Incident Command System was developed in Southern California. The paper provides a brief history of the development of communications for first responders, http://www.wireless.att.com/businesscenter/en_ US/pdf/Interoperable Com m.pdf.

[10] Air Florida Flight 90 crashed into the 14th Street Bridge and then sank into the icy Potomac River on January 13, 1982, costing 74 lives; see http://www.airdisaster.com/special/special-af90.shtml. First responders arriving on the scene from the District, Arlington County, and other counties in the capital area were unable to communicate with each other or with federal authorities who also responded.

[11] *Final Report of the Public Safety Wireless Advisory Committee,* September 11, 1996, at http://www.ntia.doc.gov/osmhome/pubsafe/pswac_al.pdf.

[12] ***Why Can't We Talk,*** February 2003: http://www.ncjrs.gov/pdffiles1/nij/204348a.pdf.

[13] *National Emergency Communications Plan,* July 2008, http://www.dhs.gov/xlibrary/ assets/national_emergency_communications_plan.pdf.

[14] ICO was created by the Ensuring Needed Help Arrives Near Callers Employing (ENHANCE) 911 Act of 2004 (P.L. 108-494), to be jointly administered by the NTIA and the National Highway Traffic Safety Administration.

[15] *A National Plan for Migrating to IP-Enabled 9-1-1 Systems,* September 2009, http://www.e-911ico.gov/NationalNG911MigrationPlan_sept2009.pdf.

[16] FCC, *Connecting America: The National Broadband Plan,* http://www.broadband. gov/ download-plan/.

[17] For example, see the FCC's Broadband Action Agenda at http://www.broadband.gov/plan/ broadband-actionagenda. html.

[18] For example, see FCC, *Fourth Notice of Proposed Rulemaking,* adopted January 25, 2011, PS Docket No. 06-229, and *Notice of Inquiry,* for Next Generation 911, adopted December 21, 2010, PS Docket No. 10-25; a summary of other actions is at http://www.fcc.gov/ pshs/ broadband.html.

[19] The continuum diagram is at http://www.safecomprogram.gov/SAFECOM/ Tools/ Continuum /continuum.htm; additional descriptions at http://www.safecomprogram.gov/ SAFECOM/ oecguidancedocuments/ continuum/default.htm.

[20] Research reports that have addressed these issues include *Sending Out an S.O.S.: Public Safety Communications Interoperability as a Collective Action Problem,* by Jerry Brito; *Communicating During Emergencies: Toward Interoperability and Effective Information Management,* by Philip J. Weiser, University of Colorado Law Legal Studies Research Paper No. 07-7, and *Fundamental Reform in Public Safety Communications Policy,* by Jon M. Peha; all published in the Federal Communications Law Journal, Vol. 59, No. 3, 2007.

[21] Long Term Evolution is a Fourth Generation (4G) mobile technology, at http://www. 4 gamericas.org/index.cfm?fuseaction=page&sectionid=249.

[22] Based on a summary of federal programs provided by SAFECOM, at http://www. safecomprogram.gov/NR/rdonlyres/132003E7-6C43-4E15-97D6-A2A4E5A2704F/0/GrantProgramsforSAFECOMWebsite.pdf.

[23] See *Appendix C.* Links to relevant SAFECOM and FEMA grant program documents are available at http://www.safecomprogram.gov/SAFECOM/grant/default.htm. Information on OEC grants is at http://www.dhs.gov/xopnbiz/grants/gc_1288707294166.shtm.

[24] Information on JAG programs is at http://www.ojp.usdoj.gov/BJA/grant/jag.html; a link to COPS programs is at http://www.cops.usdoj.gov/Default.asp?Item=46.

[25] Information on BTOP grants is reported in NTIA, *The Broadband Technology Opportunities Program,* "Overview of Grant Rewards," posted December 14, 2010, at http://www.ntia. doc.gov/reports/2010/NTIA_Report_on_BTOP_12142010.pdf.

[26] Update on 911 activities, letter to the President of the Senate from the NHTSA Administrator, December 16, 2010, http://www.911.gov/pdf/2010_%20Report_to_Congress-Biden_Letter .pdf.

[27] P.L. 108-494, §104, "Sec. 158, "(b) "(1); 118 STAT. 3987-3988.

[28] P.L. 109-171, §3004, 120 STAT. 22-23.

[29] CRS, Congressional Distribution Memorandum, "Federal Funding of State and Local Emergency Communications Projects," June 10, 2011.

[30] At least two major reports from the Government Accountability Office have documented the lack of coordination: *Homeland Security: Federal Leadership and Intergovernmental Cooperation Required to Achieve First Responder Interoperable Communications,* GAO-04-963T, July 20, 2004, at http://www.gao.gov/new.items/d04963t.pdf, and *Emergency Communications: Vulnerabilities Remain and Limited Collaboration and Monitoring Hamper Federal Efforts,* GAO-09-604, June 2009 at http://www.gao.gov/new. items/ d09604.pdf.

[31] Discussed in detail in *Appendix C.*

[32] Information about the Office of Emergency Communications is at http://www.dhs.gov/xabout/structure/gc_1189774174005.shtm.

[33] For additional information on FEMA programs, see CRS Report RL34537, *FEMA's Pre-Disaster Mitigation Program: Overview and Issues,* by Francis X. McCarthy and Natalie Keegan.

[34] 47 U.S.C. §151.

[35] See http://www.fcc.gov/pshs/about-us/.

[36] Spectrum is segmented into bands of radio frequencies and typically measured in cycles per second, or hertz. Standard abbreviations for measuring frequencies include kHz—kilohertz or thousands of hertz; MHz—megahertz, or millions of hertz; and GHz—gigahertz, or billions of hertz. The 700 MHz band includes radio frequencies from 698 MHz to 806 MHz. Current plans for the public safety broadband network would use frequencies already assigned to public safety, at 763-768MHz and 793-798 MHz.

[37] Frequencies at 769-795 MHz and 799-805 MHz.

[38] "Report from the Field: Advancing Public Safety Broadband Communications," posted December 30, 2010, at http://www2.ntia.doc.gov/node/695.

[39] See "Statewide Interoperability Coordinators" at http://www.dhs.gov/files/ programs/gc_1286986920144.shtm.

[40] See "Statewide Communication Interoperability Plans" at http://www.dhs.gov/ files/ programs/gc_1225902750156.shtm.

[41] *Establishing Governance to Achieve Statewide Communications Interoperability* at http://www.safecomprogram.gov/NR/rdonlyres/24F10648-2642-42F3-8305-B29315F833BF/0/EstablishingGovernanceGuide.pdf.

[42] Parishes and independent cities are considered counties for most statistical purposes. The Bureau of the Census identifies 3,143 counties and county-equivalents in the United States. Some states, however, have curtailed responsibilities for counties. Massachusetts, for example, has transferred governing authority from most counties to thestate or to townships.

[43] The Congressional Research Service is researching state, county, and local reports on expenditures for emergency communications equipment and has collected this information for 20 states.

[44] Hearings with testimony on the PSBN include Senate, Committee on Science, Commerce and Transportation, "Safeguarding our Future: Building a Nationwide Network for First Responders," February 16, 2011; House of Representatives, Committee on Homeland Security, "Public Safety Communications: Are the Needs of Our First Responders Being Met?," March 30, 2011, and House of Representatives, Committee on Energy and Commerce, Subcommittee on Communications and Technology, "Using Spectrum to Advance Public Safety, Promote Broadband, Create Jobs, and Reduce the Deficit," April 12, 2011; House of Representatives, Committee on Energy and Commerce, Subcommittee on Communications and Technology, "Creating an Interoperable Public Safety Network," May 24, 2011.

[45] The range of positions among public safety officials and their expectations from the federal government cited here were reiterated at a hearing on March 30, 2011. House of Representatives, Committee on Homeland Security, "Public Safety Communications: Are the Needs of Our First Responders Being Met?"

[46] House of Representatives, Committee on Energy and Commerce, Subcommittee on Communications, Technology and the Internet, "A National Interoperable Broadband Network for Public Safety: Recent Developments," September 24, 2009, written testimony

of Brian Fontes, CEO, NENA, at http://republicans.energycommerce.house.gov/Media/file/Hearings/Telecom/090924%20D-Block%20hearing/Testimony/NENA.pdf.

[47] Discussed in *Appendix B.*

[48] FCC, *Order,* P.S. Docket 06-229, released November 19, 2007, at http://fjallfoss.fcc.gov/edocs_public/attachmatch/FCC-07-199A1.pdf.

[49] FCC, *Order,* PS Docket No. 06-229, released April 23, 2010, at http://fjallfoss.fcc.gov/edocs_public/attachmatch/FCC-10-67A1.pdf.

[50] See http://www.fcc.gov/pshs/eric.html.

[51] The license covers 10 MHz at 763-768MHz and 793-798 MHz and guard bands, to mitigate interference, at 768-769MHz and 798-799 MHz.

[52] The states are Alabama, Hawaii, Iowa, Mississippi, New Jersey, New Mexico, and Oregon; the regions are Adams County-Denver Airport (CO), Bay Area (CA), Mesa-TOPAZ (AZ), San Antonio (TX), and Wisconsin Counties; the cities are Boston (MA), Charlotte (NC), Chesapeake (VA), Los Angeles (CA), New York City (NY), Pembroke Pines (FL), Seattle (WA), and Washington (DC). Of these, Alabama was not issued a spectrum lease as it was not ready to proceed.

[53] *Final Report of the Public Safety Wireless Advisory Committee,* September 11, 1996, Appendix C–ISC Final Report, p. 285, at http://www.ntia.doc.gov/osmhome /pubsafe /pswac_al.pdf.

[54] National Emergency Management Association, biennial reports available at http://www.nemaweb.org. The reports do not always provide the number of states responding to specific questions.

[55] *NEMA 2006 Biennial Report,* p. 26.

[56] *NEMA 2008 Biennial Report,* p. 17.

[57] Cyren Call Communications Corporation, in *ex parte* comments filed with the FCC on June 4, 2007, set the cumulative capital expenditure for building a public-private network at $18 billion, of which roughly a third of the cost would be for enhancements for public safety use. An estimate from Northrop-Grumman Corporation placed the cost at $30 billion, when service applications are included (statement by Mark S. Adams, Chief Architect Networks and Communications, at WCA 2007, Washington, DC, June 14, 2007). These estimates do not include the cost of radios. An estimated range of $20 billion to $40 billion for network infrastructure was discussed at a hearing held by the House of Representatives, Committee on Energy and Commerce, Subcommittee on Communications, Technology, and the Internet, "A National, Interoperable Broadband Network for Public Safety: Recent Developments," September 24, 2009.

[58] Additional information is provided in *Appendix B,* "Proposals for Spectrum Assignment." 59 See *Appendix B.* The D Block frequencies at 758-763 MHz and 788-793 MHz are adjacent to the PSBL.

[60] FCC, *A Broadband Network Cost Model: A Basis for Public Funding Essential to Bringing Nationwide Interoperable Communications to America's First Responders,* OBI Technical Paper No. 2, May 2010, http://download.broadband.gov/plan/fcc-omnibus-broadband-initiative-%28obi%29-technical-paper-broadbandnetwork-cost-model-basis-for-public-funding-essential-to-bringing-nationwide-interoperable-communications-toamericas-first-responders.pdf.

[61] Ibid., Exhibit 6.

[62] Ibid., Exhibit 7.

[63] NPSTC is a federation of associations whose mission is to improve public safety communications and interoperability through collaborative leadership, http://www.npstc.org/index.jsp.

[64] NPSTC Position Paper, http://www.npstc.org/documents/NPSTC%20Position%20BB% 20 Network%2020110310.pdf.

[65] NPSTC, National Broadband Network Conceptual Design, March 3, 2011, http://www.npstc.org/documents/20110301%20EPC-AN%20Drawings%20FINAL%20w%20notes.pdf.

[66] Quotes are from NPSTC Position Paper, *op.cit.*

[67] "Quantifying the Costs of a Nationwide Public Safety Wireless Network," by Ryan Hallahan and Jon M. Peha,authors' version at http://www.andrew.cmu.edu/user/rhallaha/ papers/quantifying_costs_of_PS_network.pdf. Also published in Telecommunications Policy, October 2010.

[68] "The Benefits of Transitioning to a Nationwide Wireless Broadband Network for Public Safety," The White House, released by the Office of the Vice President, June 2011, http://www.whitehouse.gov/sites/default/files/uploads/publicsafetyreport.pdf.

[69] U.S. Department of Transportation, Intelligent Transportation Systems, "Next Generation 9-1-1 System Initiative: Final Analysis of Cost, Value, and Risk," March 8, 2009, pp 57-58 and 62-64, at http://www.its.dot.gov/ng911/pdf/USDOT_NG911_4-A2_FINAL_FinalCost ValueRiskAnalysis_v1-0.pdf.

[70] Department of Homeland Security, S&T Snapshots, Command, Control, and Interoperability, "The Beginning of the End of the Single-Band Radio for Public Safety," October 20, 2009, at http://www.dhs.gov/files/programs/gc_1258141690101.shtm. In testimony on March 30, 2011, Mr. Gregory Simay, At-Large Director, Los Angeles Regional Interoperable Communication System estimated the cost of a broadband network radio at over $7,000, House of Representatives, Committee on Homeland Security, "Public Safety Communications: Are the Needs of Our First Responders Being Met?"

[71] Obstacles to developing an integrated voice-data device for first responders are discussed in "LTE Support for Mission Critical Voice for Public Safety," by Andrew M. Seybold, Public Safety Advocate e-newsletter, June 19, 2011, at http://andrewseybold.com/2532-lte-support-for-mission-critical-voice-for-public-safety.

[72] Dated July 20, 2010, http://energycommerce.house.gov/documents/ 0100726/ Letter. FCC. 07.26.2010.pdf.

[73] Dated April 11, 2011, http://republicans.energycommerce.house.gov/Media/file/Letters/ 112th /042011FCC.pdf.

[74] Dated May 19, 2011. The FCC has allowed some public safety agencies to begin building broadband networks by waiving some of their existing rules concerning spectrum allocation. FCC, *Order,* released December 10, 2010, PS Docket No. 06-229.

[75] FCC, Public Notice, "Public Safety and Homeland Security Bureau Seeks Comment on Increasing Public Safety Interoperability by Promoting Competition for Public Safety Communications Technology," PS Docket No. 10-168, released August 19, 2010, http://fjallfoss.fcc.gov/edocs_public/attachmatch/DA-10-1556A1.pdf.

[76] Estimate reported in *Progress Report to Congress on Emergency Communications,* 2011, DHS, Office of Emergency Communications, p. 7.

[77] S. 28 (Rockefeller), S. 1040 (Lieberman), S. 911, as amended (Rockefeller), and H.R. 607 (King).

[78] 47 U.S.C. 309 (j) (8) (A).

[79] 47 U.S.C. 309 (j) (7) (B).

[80] 2 MHz has also been assigned to the PSBL as guard bands, to protect against radio traffic interference.

[81] According to its website, the Annenberg Innovation Lab, based within the Annenberg School for Communication and Journalism at the University of Southern California, fosters innovation through collaborative projects with other University departments, public institutions and the private sector; see http://www.annenberglab.com/about.

[82] As described in an e-mail to Congressional Research Service on April 18, 2011. The team consists of computer science students from the Viterbi School of Engineering at the University of Southern California.

[83] Ultrawide band provides broadband capacity but is only effective for short-range communications, for example, in deploying bomb-detection robots or monitoring sensors and cameras in a burning building. In the commercial sector it is widely used to provide wireless connectivity in homes.

[84] Also known as 3GPP Release 10, see http://www.3gpp.org/LTE-Advanced.

[85] *Department of Defense Global Information Grid Architectural Vision; vision for a net-centric, service-oriented, DoD enterprise,* prepared by the DOD CIO, June 2007, http://cio-nii.defense.gov/docs/GIGArchVision.pdf.

[86] Ibid., Figure 8, p. 22.

[87] FEMA and the FCC have announced the introduction of the Personal Localized Alerting Network (PLAN) that can deliver geo-targeted text alerts to enabled smart phones. PLAN is the implementation of the Commercial Mobile Alert Service as required by the Warning, Alert, and Response Network Act (WARN Act), P.L. 109-347, §603 (a). FCC Fact Sheet, http://transition.fcc.gov/cgb/consumerfacts/cmas.pdf.

[88] U.S. Army Research, Development and Engineering Command, U.S. Army Communication-Electronics Research, Development and Engineering Center, Multi-Access Cellular Extension Demonstration Project.

[89] Mesh networks operate over linked radio nodes facilitated in part by a mesh router. The nodes can form ad hoc networks, as needed, and operate without cell towers. The mesh network serving the City of Minneapolis is widely credited with expediting response and recovery, and saving lives after a bridge spanning the Mississippi collapsed into the river on August 1, 2007. More recently, Japan's mesh networks helped maintain communications after the earthquake and subsequent tsunami in mid-March 2011.

[90] House of Representatives, Committee on Armed Services, Subcommittee on Emerging Threats and Capabilities, "Department of Defense Science and Technology Budget Request," March 1, 2011, written statement of Dr. Marilyn Freeman, Deputy Assistant Secretary of the Army for Research and Technology.

[91] "Can You Hear Me Now? Vice Chief of Staff Attends Army Communications Demonstration," by Sharon Rushen, RDECOM CERDEC Public Affairs, March 10, 2011, at http://www.cerdec.army.mil/news/comms_demo.asp.

[92] Additional information is provided in Congressional Distribution Memoranda, "Spectrum Policy and Deficit Reduction: Administration-Sponsored Bills," December 7, 2011 and "Public Safety and Spectrum Policy Provision in H.R. 3630 and Other Bills," December 14, 2011, available on request to the author.

[93] Frequencies assigned to narrowband are 769-775 MHz and 799-805 MHz. There are also two assignments to public safety of 1 MHz each that are designated as guard bands to prevent interference from adjacent channels.

[94] S. 28, §107.

[95] S. 28, §105 (f).

[96] S. 28, §202 (a), (b) and (d).
[97] S. 28, §105 (b) (1) (G).
[98] S. 28, §209.
[99] S. 28, §101.
[100] S. 28, §102 (b) (3).
[101] S. 28, §105 (b) (1) (D).
[102] S. 28, §105 (c).
[103] S. 28, §105 (d).
[104] S. 28, §207 and §208.
[105] S. 28, §204.
[106] S. 28, §104 (a) (5).
[107] S. 28, §201 (a).
[108] S. 28, §204 (b).
[109] S. 28, §210.
[110] S. 28, §102 (a).
[111] S. 28, §102 (b).
[112] S. 28, §104.
[113] S. 28, §106.
[114] S. 28, §103.
[115] S. 28, §205.
[116] S. 28, §105.
[117] S. 28, §105 (e).
[118] S. 28, §206.
[119] S. 28, Sec, 204.
[120] H.R. 2482, §202.
[121] H.R. 2482, §203.
[122] H.R. 2482, §401 (a).
[123] H.R. 2482, §401 (b) (4).
[124] H.R. 2482, §401 (b) (3).
[125] H.R. 2482, §208.
[126] H.R. 2482, §101.
[127] H.R. 2482, §102.
[128] H.R. 2482, §401 (b) (5).
[129] H.R. 2482, §401 (b) (6).
[130] H.R. 2482, §301.
[131] H.R. 2482, §303.
[132] H.R. 2482, §302.
[133] S. 1549, §284.
[134] S. 1549, §285.
[135] S. 1549, §288.
[136] S. 1459, §297 (a).
[137] S. 1459, §297 (b).
[138] S. 1459, §297 (b) (4).
[139] S. 1459, §297 (b) (3).
[140] S. 1459, §290.
[141] S. 1549 (§281).
[142] S. 1549 (§282).
[143] S. 1549, §297 (b) (5).

[144] S. 1459, §275.
[145] S. 1459, §273.
[146] S. 1459, §297 (b) (2).
[147] S. 1459, §277 (a).
[148] S. 1459, §277 (b).
[149] S. 1040, §203 (a).
[150] S. 1040, §204.
[151] S. 1040, §205.
[152] S. 1040, §106.
[153] S. 1040, §106 (1) and (2).
[154] S. 1040, §105 (a).
[155] S. 1040, §105 (b).
[156] S. 1040, §203 (b) (1) and (2).
[157] S. 1040, §203 (e).
[158] S. 1040, §202 (b).
[159] S. 1040, §202 (b) (1).
[160] S. 1040, §202 (b) (2).
[161] S. 1040, §201 (b) (3).
[162] S. 1040, §202 (c).
[163] S. 1040, §103 (a) (3) (C).
[164] S. 1040, §102.
[165] S. 1040, §103 (a) (1).
[166] S. 1040, §103 (a) (3) (A).
[167] S. 1040, §207 (c).
[168] S. 1040, §103 (a) (4).
[169] S. 1040, §207 (b)
[170] S. 1040, §207 (d).
[171] S. 1040, §207 (d) (2).
[172] S. 1040, §207 (b) (2).
[173] S. 1040, §207 (b) (3).
[174] S. 1040, §207 (a).
[175] S. 1040, §103 (a) (2).
[176] S. 1040, §104.
[177] S. 1040, §208.
[178] S. 1040, §206 (c).
[179] S. 1040, §206 (a).
[180] S. 1040, §206 (b).
[181] H.R. 607, §104.
[182] **H.R. 607, §102 (b) '(b) '(2) '(C).**
[183] H.R. 607, §205.
[184] **H.R. 607, §102 (b) '(b) '(2) '(A).**
[185] H.R. 607, §202 (a) (1) (A); §203; and §204.
[186] H.R. 607, §202 (b).
[187] H.R. 607, §202 (b) (1) and (2).
[188] H.R. 607, §202 (e).
[189] **H.R. 607, §102 (a) and (b) '(b) (2) (A).**
[190] H.R. 607, §103 (a).
[191] H.R. 607, §207 (a).

[192] H.R. 607, §207 (c).
[193] **H.R. 607, §102 (b) '(b) '(2) '(B).**
[194] H.R. 607, §103 (a).
[195] H.R. 607, §103 (b).
[196] H.R. 607, §210.
[197] H.R. 607, §208.
[198] H.R. 607, §207 (b).
[199] H.R. 607, Dec. 207 (d).
[200] The FCC manages spectrum access for commercial and other nonfederal uses. The National Telecommunications and Information Administration is responsible for overseeing spectrum used by federal agencies.
[201] FCC, *Second Report and Order,* July 31, 2007, WT Docket No. 96-86 and *Third Further Notice of Proposed Rulemaking,* released September 25, 2008, PS Docket No 06-229, http://fjallfoss.fcc.gov/edocs_public/attachmatch/FCC-08-230A1.pdf. Additional comments have been sought through other mechanisms and the docket remains open.
[202] P.L. 111-5, Division B, Title VI, §6001 (k) (1), 123 STAT. 515.
[203] FCC, NBP Public Notice # 8, *Additional Comment Sought on Public Safety, Homeland Security, and Cybersecurity Elements of National Broadband Plan,* DA 09-2133, released September 08, 2009, GN Docket Nos. 09-47, 09-51, 09-137, at http://hraunfoss.fcc.gov/edocs_public/attachmatch/DA-09-2133A1.pdf.
[204] PS Docket No. 06-229.
[205] Proposals for building a new network and assigning spectrum were the topic of a hearing held by the House Committee on Energy and Commerce, Subcommittee on Communications, Technology, and the Internet, "A National, Interoperable Broadband Network for Public Safety: Recent Developments," September 24, 2009.
[206] 47 U.S.C. 309 (j) (8) (A).
[207] Office of Management and Budget, *The Federal Budget: Fiscal Year 2012,* Appendix: Other Federal Agencies, Federal Communications Commission (p. 1227 *et seq.).*
[208] 47 U.S.C. §309 (j) (14).
[209] Additional information at http://www.fcc.gov/pshs/public-safety-spectrum/700-MHz/regional-p lanning.html.
[210] Notably, P.L. 107-296, §232, 116 STAT. 2159 and §502, 116 STAT. 2213.
[211] "Homeland Security Starting Over With SAFECOM," *Government Computer News,* June 9, 2003.
[212] P.L. 108-458, Title VII, Subtitle C, §7303 (a) (2), 118 STAT. 3843-3844.
[213] *The 9/11 Commission Report: Final Report of the National Commission on Terrorist Attacks Upon the United States,* Washington: GPO, 2004, p. 397.
[214] P.L. 108-458, Title VII, Subtitle C, §7303, 118 STAT. 3843 *et seq.*
[215] P.L. 108-458, Title VII, Subtitle C, §7304, 118 STAT. 3847-3848.
[216] P.L. 108-458, Title VII, Subtitle C, §7303 118 STAT. 3843 *et seq.*
[217] P.L. 109-171, §3002 (a) (1) (B). The deadline was later extended through June 12, 2009, by the DTV Delay Act, P.L. 111-4, §2 (a) (1).
[218] P.L. 109-171, §3004, 120 STAT. 22-23.
[219] P.L. 109-171, §3006, 120 STAT. 24-25.
[220] Congressional Budget Office Cost Estimate, S. 1932, Deficit Reduction Act of 2005, January 27, 2006, p. 21, http://www.cbo.gov/showdoc.cfm?index=7028&sequence=0.
[221] P.L. 109-459, §2.

[222] P.L. 110-53, Implementing Recommendations of the 9/11 Commission Act of 2007, Title XXII, §2201, 121 STAT. 537 *et seq.*

[223] MOU at http://www.ntia.doc.gov/otiahome/psic/PSICMOU_Executed_2-16-2007.pdf.

[224] Testimony of Corey Gruber, Acting Assistant Secretary for Grants and Planning, Office of Grants and Training, Department of Homeland Security at hearing on "Public Safety Interoperable Communications Grants: Are the Departments of Homeland Security and Commerce Effectively Coordinating to Meet our Nation's Emergency Communications Needs?" House of Representatives, Homeland Security Committee, Subcommittee on Emergency Communications, Preparedness, and Response, March 14, 2007.

[225] Press releases at http://www.dhs.gov/xnews/releases/pr_1184783934669.shtm and http://www.ntia.doc.gov/ntiahome/press/2007/PSIC_071807.pdf.

[226] See http://www.dhs.gov/xgovt/grants/gc_1184774852768.shtm. The NTIA website main page has a section devoted to PSIC at http://www.ntia.doc.gov.

[227] FEMA, Grants Program Directorate, Information Bulletin No. 337, November 23, 2009, at http://www.ntia.doc.gov/psic/IJ/wv.pdf.

[228] NTIA, *Second Annual Assessment of the Public Safety Interoperable Communications Grant Program,* Final Report No. OIG-11-001-A, October 7, 2010, at http://www.oig.doc.gov/oig/reports/2010/OIG-11-001-A.pdf.

[229] See http://www.oig.doc.gov/Pages/National-Telecommunications-and-Information-Administration.aspx?YearStart=01/01/2010&YearEnd=12/31/2010 and http://www.oig.doc.gov/Pages/National-Telecommunications-and-Information-Administration.aspx?YearStart=01/01/2009&YearEnd=12/31/2009. The states are Texas, Florida, Massachusetts, California, Pennsylvania, Louisiana, Nevada, New York, and Arkansas.

[230] Information on the FEMA reorganization is provided in CRS Report RL33729, *Federal Emergency Management Policy Changes After Hurricane Katrina: A Summary of Statutory Provisions,* coordinated by Keith Bea.

[231] P.L. 109-295, Title VI, §671(b), 'Title XVIII, 'Sec. 1805, 120 STAT. 1439.

[232] P.L. 109-295, Title VI, §671(b), 'Title XVIII, 'Sec. 1803, 120 STAT. 1437-1438.

[233] DHS, National Emergency Communications Plan, July 2008, http://www.dhs.gov/xlibrary/assets/national_emergency_communications_plan.pdf.

[234] "Our vision was developed at the 2003 SAFECOM/AGILE Joint Program Planning Meeting in San Diego, CA.", Emergency Response Council, *Agreements on a Nationwide Plan for Interoperable Communications,* Summer 2007, footnote 1.

[235] *Op. cit., Agreements on a Nationwide Plan for Interoperable Communications.*

[236] P.L. 109-295, Title VI, §671(b), "Title XVIII, "Sec. 1805, 120 STAT. 1439.

[237] Presentation by Brian Carney, National RECC Coordinator, Disaster Operations Directorate, Federal Emergency Management Agency, Department of Homeland Security, "Regional Emergency Communications Coordination Working Groups (RECCWG)" at National Public Safety Telecommunications Council (NPSTC), Seattle, Washington, September 15, 2008, ttp://www.npstc.org/meetings/GB_Carney_RECC%20Briefing_090208_NPSTC.ppt.

[238] DHS, National Emergency Communications Plan, July 2008: http://www.dhs.gov/xlibrary/assets/national_emergency_communications_plan.pdf.

[239] National Emergency Communications Plan, "Executive Summary," p. ES-1.

[240] Ibid., "Executive Summary," p. ES-2.

In: Emergency Communications ISBN: 978-1- 62081-140-5
Editors: D. H. Martinez and A. E. Young © 2012 Nova Science Publishers, Inc.

Chapter 2

PUBLIC SAFETY COMMUNICATIONS AND SPECTRUM RESOURCES: POLICY ISSUES FOR CONGRESS*

Linda K. Moore

SUMMARY

Effective emergency response is dependent on wireless communications. To minimize communications failures during and after a crisis requires ongoing improvements in emergency communications capacity and capability. The availability of radio frequency spectrum is considered essential to developing a modern, interoperable communications network for public safety. Also critical are (1) building the network to use this spectrum and (2) developing and deploying the radios to the new standards required for mobile broadband. Beyond recognition of these common needs and goals, opinions diverge on such issues as how much spectrum should be made available for public safety broadband communications, how communications networks should be configured, who should own them, who should build them, who should operate them, who should be allowed to use them, and how they might be paid for.

Three bills that would increase the amount of radio frequency spectrum assigned for public safety use have been introduced. The bills

* This is an edited, reformatted and augmented version of a Congressional Research Service publication, CRS Report for Congress R40859, from www.crs.gov, prepared for Members and Committees of Congress, dated September 1, 2010.

would require that the Federal Communications Commission (FCC) transfer a spectrum license intended for commercial use, known as the D Block, to the license-holder for adjacent frequencies already assigned to public safety, known as the Public Safety Broadband License. The Broadband for First Responders Act of 2010 (H.R. 5081, Representative King) deals primarily with reassignment of the D Block. Two Senate bills contain similar provisions for spectrum assignment and would add a number of new provisions, including using the proceeds of future spectrum auctions to fund the needed network (S. 3625, Senator Lieberman and S. 3756, Senator Rockefeller). The development of public safety radios for broadband would be expedited by companion bills H.R. 5907 (Representative Harman) and S. 3731 (Senator Warner). Public safety operations would benefit from the radio-development initiative regardless of the eventual assignment of the D Block. Congress may consider additional legislation or oversight to meet desired levels of emergency communications performance.

Among the actions that Congress might take, those dealing with governance and funding are often cited by public safety officials and others as the areas most in need of its consideration. Many have recommended that, for the proposed broadband network projects to go forward on a sustainable footing, funding sources need to be identified for investment and operating expenses over the long term. To ensure the resources are wisely used, some analysts point to the primacy of putting in place a well-grounded but flexible governance structure. The debate on spectrum assignment has in recent months dominated the attention of Congress and other policy makers. Meanwhile, several states and urban areas including the San Francisco Bay Area, Boston, and the State of Mississippi have submitted detailed plans for building the nation's first broadband networks that might serve as a practical framework for evaluating policy options. These plans, developed according to FCC requirements, share many common features. Notable from a policy point of view are several recommendations that provide a common theme in these early submissions. These may be summarized as: (1) sufficient funding is essential; (2) networks that either cover an area designated as eligible for Urban Area Security Initiative programs, or cover a regional area—that is, large and/or densely populated areas—are more efficient to build, operate, and govern; (3) several critical technologies and standards, such as for radios, must be developed before the networks can be fully effective; (4) some form of governing sur-structure must be in place to assure uniformity of core operations while allowing for local customization of public safety applications; and (5) collaboration with commercial partners is important for mustering all the skills and knowledge resources needed for developing the leading-edge broadband networks that are the goals of the submitted plans.

THE ISSUES BEFORE CONGRESS

Since September 11, 2001, Congress has passed several significant pieces of legislation intended to help surmount failures in public safety radio communications such as (1) insufficient interoperability among radio systems, a problem that hampered rescue efforts on and after September 11; and (2) insufficiently robust networks, a shortcoming revealed after Hurricane Katrina struck in August 2005. To achieve a higher standard of communications performance might require, among other elements, improvements in communications capacity and quality. Increased capacity is achievable through a number of means. Increasing the amount of radio frequencies available for public safety use is one solution for adding capacity. Building additional infrastructure to use existing airwaves more effectively is another solution, as is investment in more spectrum-efficient technologies. Sharing networks also can provide additional capacity for operations. All of these measures have been proposed for improving public safety communications, with different groups voicing preferences for one means over another.

Many representative of the public safety community have argued that additional spectrum assignments are needed to meet the future needs of emergency communications, while the Federal Communications Commission (FCC) has presented an action plan that would develop capacity through investing in network infrastructure, public-private sharing of development costs for efficient radios, and creating a regulatory regime that would allow public safety and commercial users to share infrastructure. All of the measures under consideration by the FCC or proposed by public safety agencies would require substantial funding—many billions of dollars— of which some is expected to come from the federal government.

Three bills have been introduced that would require the FCC to assign additional spectrum, known as the D Block, for a public safety broadband network and take steps to ensure construction of an interoperable network.[1] A draft discussion bill was released in June that would support the FCC's plans for using spectrum and developing infrastructure, funded in part by auction proceeds that would include the sale of the D Block.[2] These bills deal lightly with the question of governance, for the most part making adjustments to the regulatory and oversight responsibilities of the FCC.

The Next Generation Public Safety Device Act of 2010 (H.R. 5907, Representative Harman) and its companion bill (S. 3731, Senator Warner) are more narrowly focused on a critical initial step on the long road to assuring that a nationwide, interoperable network is put in place for public safety communications: the radios. The FCC has linked the auction of the D Block to the development of the needed public safety features for broadband radios that would operate on the D Block network frequencies, the Public Safety Broadband License frequencies, and perhaps other frequencies within the band. In the FCC's announced plans, the D Block license-holder(s) would take the lead in, and assume the costs of, developing appropriate hardware and software for radios used by public safety. The FCC, in the NBP, stated that details of how these requirements would be determined and enforced would likely be decided as part of the comment process that would lead up to the auction, with development work starting after a successfully completed auction. The Public Safety Device Act would begin the process of development almost immediately and would encourage innovation and competitive pricing through a technology competition administered by the National Telecommunications and Information Administration (NTIA).

DEBATE OVER SPECTRUM RESOURCES: THE D BLOCK

Congress last addressed the public safety community's need for spectrum capacity by mandating the release of 24 MHz[3] of frequencies that were originally designated for public safety use in the late 1990s.[4] This crucial resource, part of the 700 MHz band,[5] remained largely unavailable as long as its airwaves were used for analog television transmissions. By providing a deadline for the transition from analog to digital television, Congress ensured that valuable radio frequency spectrum would be released by 2009.[6]

The assignment of one set of frequencies in the 700 MHz band, referred to as the D Block, has been widely debated. The D Block was slated for auction in 2008 along with other available frequencies identified in the Deficit Reduction Act of 2005.[7] In compliance with instructions from Congress to auction all unallocated spectrum in this band, the FCC conducted an auction, which concluded on March 18, 2008. As part of its preparation for the auction (Auction 73), the FCC sought to increase the amount of spectrum available to public safety users in the 700 MHz band. The FCC proposed to assign 10 MHz—part of the original 24 MHz designated for public safety use—to a Public Safety Broadband Licensee specifically for public safety broadband communications. Of the balance, 12 MHz were designated for mission critical voice communications on narrowband networks and 2 MHz were set aside as a guard band to protect against interference. A section of the 700 MHz band plan, showing the location of public safety licenses and the D Block, is provided in *Appendix B*.

In the FCC plan for Auction 73, the Public Safety Broadband License (PBSL) was to have been matched with a commercial license of 10 MHz, known as the D Block. The D Block was to be auctioned under rules that would require the creation of a public-private partnership to develop the two 10-MHz assignments as a single broadband network, available to both public safety users and commercial customers. The D Block license was offered for sale in 2008 but did not find a buyer. The FCC then set about the task of writing new rules for a reauction of the D Block.[8]

FCC's Announced Plans for the D Block

The FCC subsequently decided to auction the D Block for commercial use with conditions deemed beneficial for public safety users, such as assumption by the license-holder of the cost of developing mobile devices, and guarantees that public safety networks would have roaming and priority access rights to the D Block network. The decision was announced in the National Broadband Plan (NBP),[9] released March 16, 2010. The NBP proposed several actions to be taken to facilitate development of a national wireless broadband network for public safety use. [10] Public safety needs, such as developing standards and establishing procedures, would be addressed through a newly established Emergency Response Interoperability Center (ERIC).[11]

Legislation to Assign the D Block to Public Safety

The Broadband for First Responders Act of 2010 (H.R. 5081, Representative King) would amend the Communications Act of 1934 by requiring the FCC to allocate.the D Block for public safety services. The bill would require the FCC to establish rules to encourage the rapid deployment of an interoperable national wireless broadband network, and to allow public safety license-holders to share spectrum with other entities, as long as requirements for roaming and priority access were met. The First Responders Protection Act of 2010 (S. 3625, Lieberman) and the Public Safety Spectrum and Wireless Innovation Act (S. 3756, Rockefeller) would require similar rule-making procedures, and requirements.

Legislation in Support of a Public Safety Network without the D Block

The discussion draft of the Public Safety Broadband Act of 2010 includes the presumption that the D Block will be auctioned, in that it provides that proceeds from its auction be applied to the construction and operation costs of public safety broadband networks. The draft bill would permit sharing of spectrum designated for broadband networks between public safety and other entities. It would also direct the FCC to allow flexible use of other frequencies in the 700 MHz band designated for public safety.

COMMUNICATIONS INFRASTRUCTURE AND GOVERNANCE

The First Responders Protection Act of 2010, the Broadband for First Responders Act of 2010, the Public Safety Spectrum and Wireless Innovation Act, the discussion draft of the Public Safety Broadband Act of 2010, the Next Generation Public Safety Device Act of 2010, and several inter-connected initiatives of the FCC address aspects of how to plan, build, and fund a national network for public safety communications.

Public Safety Broadband Network Requirements

Developments in mobile broadband communications are changing the public safety community's expectations about how to best use the 700 MHz airwaves allocated for their use. Public safety representatives have argued that this spectrum should be used for a wireless network customized to meet needs that they have indentified.

Arguments in favor of building a network exclusively for public safety revolve around the shortcomings of current commercial wireless services such as poor availability, inadequate coverage in rural areas, lack of security features, and absence of priority access.

Network infrastructure requirements for public safety communications that are frequently discussed include[12]

- Broadband applications should facilitate emergency response by providing data and images, including video.[13]

- The network should cover all areas of the United States, ensuring service to meet a public safety emergency anywhere.
- Broadband services should include voice communications as a back up to mission critical voice channels on other frequencies and offer the same features such as push-to-talk and one-to-one or one-to-many connectivity.
- Network software should provide traffic management services such as prioritizing service. If multiple networks were built separately and then linked together, interoperability[14] and nationwide roaming[15] would need to be ensured.
- Radio software should provide mobile broadband applications designed for public safety. In particular, radio chipsets need to be developed for wireless devices that can connect to a Long Term Evolution (LTE) network.
- Radio software should support encryption and authentication.
- Cell towers in the network should be strengthened against natural hazards and furnished with back-up power supplies that can outlast extended power outages.
- Robust backhaul should be ensured. Backhaul typically refers to connectivity between access points like cell towers and high capacity, landline communications networks. Backhaul is an essential component of wireless network infrastructure.

FCC's Proposals for Communications Infrastructure

In the NBP, the FCC made these key recommendations for promoting public safety wireless broadband communications.[16]

- Create an administrative system that ensures access to sufficient capacity on a day-to-day and emergency basis.
- Ensure there is a mechanism in place to promote interoperability and operability of the network.
- Establish a funding mechanism to ensure the network is deployed throughout the United States and has necessary coverage, resiliency, and redundancy.
- Conform existing programs to operate with the public safety broadband network.

The FCC has recommended that the public safety community leverage the availability of commercial technologies and networks to assure system-wide capacity and has encouraged partnerships and administrative agreements with commercial operators and others.

Emergency Response Interoperability Center

The FCC would address public safety needs such as developing standards and establishing procedures through the newly established Emergency Response Interoperability Center (ERIC). ERIC was established within the FCC Public Safety and Homeland Security Bureau, in April 2010.[17] It is intended for ERIC to work closely with the Public Safety Communications Research program, jointly managed by the National Institute of Standards and Technology (NIST) and the NTIA, to develop and test the technological solutions needed for public safety broadband communications.[18] The Department of Homeland Security is to participate in the areas of public safety outreach and technical assistance, as well as best practices development, through its Office of Emergency Communications. ERIC has been tasked with implementing standards for national interoperability and developing technical and operational procedures for the public safety wireless broadband network in the 700 MHz band. In the future, ERIC may perform similar functions for other public safety communications systems.

Within the 700 MHz band, ERIC might use the regulatory powers of the FCC to require the cooperation of commercial wireless operators in establishing roaming rights and access rules between the public safety broadband network and other networks built to use the 700 MHz frequencies. In particular, the FCC's powers to write rules for spectrum license auctions and set service rules for auction winners are to be brought to bear on the winner or winners of licenses in the D Block. and it has anticipated that it will be able to negotiate roaming and priority access across the 700 MHz band. Its authority to enforce access requirements is uncertain, however, and might be successfully challenged in court.

Requirements for Conditional Build-Outs

Some states and localities have petitioned the FCC to allow them to use frequencies from the 10 MHz assigned to the PSBL for their own public safety networks. Plans would be developed based on local and regional needs, with anticipated funding from sources such as existing programs, partnerships with commercial providers, and federal grants. The FCC has therefore adopted an order to provide the framework for nationwide interoperability and mobile broadband and grant waivers to public safety entities that meet its requirements.[19] Providing an "Interoperability Showing" that includes evidence of funding is among the conditions established by the FCC. Fifteen recipients of conditional waivers had submitted Interoperability Showings by August 17, 2010.[20] Only one, from the San Francisco Bay Area, was accepted for review. The remainder were returned for additional work because not all the required elements had been included.[21] Nonetheless, a review of the submission seems to suggest that the materials for building a cohesive national policy are embedded in these plans.

ERIC will play a lead role in approving and coordinating the technical aspects of the waiver requests. Technical components of the waiver requests include specifications for system architecture, required applications, and network operations, administration and maintenance. System requirements to support interoperability must include radio access network and core network architectures. Plans for supporting roaming, priority access, Quality of Service (QoS), and security are required. Specifications must be provided regarding the devices planned for use on the network, including information on type (form factor), operational specifications, and spectrum coverage.[22]

Legislative Proposals for Communications Infrastructure

The First Responders Protection Act of 2010, the Broadband for First Responders Act of 2010, and the Public Safety Spectrum and Wireless Innovation Act would require the FCC to establish rules for the construction and operation of a wireless public safety broadband network. The requirements would cover interoperability, roaming, priority access, network survivability, and cybersecurity. The FCC would also be required to develop a statement of requirements for standards that would take into account: commercial availability of technologies; licensing terms; adaptability; transmission priority; security; and other considerations, as appropriate.

The discussion draft of the Public Safety Broadband Act of 2010 eliminates requirements for roaming privileges but otherwise would make similar requirements. It would direct the FCC to "take all actions necessary" to develop and implement technical standards and rules for a nationwide public safety interoperable broadband network that would include user authentication and encryption. This and other provisions in the bill would support the FCC's regulatory authority to mandate sharing of 700 MHz infrastructure. The FCC would be required to establish "an appropriate rule, or set of rules" to ensure interoperability, taking into account: commercial availability of technologies; licensing terms; adaptability; transmission priority; and security. The Public Safety Spectrum and Wireless Innovation Act also would address the FCC's need to require roaming, but in more limited terms. It would provide that the FCC "may adopt rules, if necessary in the public interest, to improve the ability of public safety networks to roam onto commercial networks and to gain priority access ..." if certain provisions are met.

The First Responders Protection Act of 2010 does not include any provisions that support the FCC's position to mandate that commercial networks provide access for roaming or make other accommodations to public safety users. These provisions will likely be part of separate contractual agreements between public safety network users and commercial network owners as part of negotiations among multiple players.

The First Responders Protection Act of 2010, the Public Safety Spectrum and Wireless Innovation Act, and the discussion draft provide funding mechanisms for construction and operation of the wireless broadband network. Projects eligible for funding that are mentioned in the bills include "construction of a new public safety interoperable broadband network using commercial infrastructure or public safety infrastructure, or both, in the 700 MHz band" and "improvement of the existing commercial networks and construction of new infrastructure to meet public safety requirements...." There is no provision in any of the bills for funds to cover the cost of development and testing of new radio technologies that allow public safety broadband radios to operate on the new networks.

The Public Safety Spectrum and Wireless Innovation Act has a number of additional provisions related to the requirements for network construction. For example, there are several provisions that are intended to assure that rural areas are given equality with urban areas in planning and construction. The bill also would require that interoperability planning include "integration with 9-1-1 call centers."

COMMUNICATIONS INFRASTRUCTURE AND RADIOS

Additional development work is needed to advance from the planning stages to testing and deployment of mobile devices that operate on the broadband network. The Next Generation Public Safety Device Act of 2010 would address the issue of developing mobile devices for broadband communications on frequencies assigned for broadband use, and other frequencies, where feasible. One of the constraints in constructing mobile devices is that it becomes increasingly difficult and expensive for radios to operate on multiple bands.

Public safety officials, commercial network experts, and the FCC are generally in agreement that Long Term Evolution (LTE) technology should be required for the new broadband network.The term profile is generally used in referring to the range of technical specifications needed for mobile devices using LTE technology to operate on a designated network. The primary group coordinating standards-setting for LTE[23] has established four profiles for commercial bands using LTE in the 700 MHZ band: Band 12, Band 13, Band 14, and Band 17. Band 14 includes the D Block and can include the public safety frequencies assigned to broadband and possibly the frequencies assigned to narrowband as well. The LTE profile for Band 14 needs to be modified to support public safety requirements. Part of the challenge for ERIC and network developers participating in the early-build-out program will be to establish a profile for public safety requirements that can be developed in conjunction with the Band 14 profile for the D Block and, possibly, other LTE bands at 700 MHz.

FCC Proposals for Radio Development

In addition to cooperation for sharing network resources, the FCC has anticipated that the D Block owner or owners will lead, and fund, the development costs of the air interface that will operate within the band comprised of the Public Safety Broadband License and the D Block. LTE has been specified by the FCC as the network technology for these frequencies. The FCC has also assumed that the other networks at 700 MHz will use LTE or a compatible fourth-generation (4G) technology.

One of the expectations is that ERIC will be able to guide the development of standards for crucial radio components, with the participation of commercial providers and public safety representatives. The participation of commercial carriers in developing and deploying, for example, a common radio interface, is expected to put the cost of public safety radios in the same price range as commercial high-end mobile devices ($500). By contrast, interoperable radios for the narrowband networks at 700 MHz cost $3,000 and up, each.

In a letter responding to an inquiry from Representative Henry A. Waxman, Chairman of the House Committee on Energy and Commerce, the

FCC provided a summary of its findings and views regarding market competition for narrowband radios.[24] The general conclusions of the letter were that proprietary technologies had hampered the effective development of public safety radios and curtailed interoperability.

Legislative Proposals for Radio Development

The Next Generation Public Safety Device Act of 2010 would require the NTIA, with the agreement of an appropriate working group, to establish requirements for Radio over Internet Protocol Devices (RoIP) and to award grants for the development of these devices. Requirements to compete in the first stage of a three-stage competition include: the identification of specific communication needs of public safety personnel and any corresponding device features that would meet those needs; and ensuring rapid, reliable, comprehensible and interoperable communications. Requirements that must be met in the final stage include submission of a plan for commercial production. In considering the winners at each stage of the process, the NTIA is required to give consideration to additional factors such as encouragement of competition; estimated cost; and the potential to use other portions of the 700 MHz band licensed for public safety use.

FUNDING

At the time of the attempted auction of the D Block, the cost of building the mobile broadband network under the public-private partnership proposed by the FCC was estimated at from $18 billion to as much as $40 billion.[25] These projected costs did not include radios.

FCC's Proposals for Funding Infrastructure

In the NBP, the FCC has recommended that a grant program be established to ensure that needed infrastructure is fully deployed.[26] It has recommended that the grants program be administered by a single agency and only be applied to projects that comply with requirements set by ERIC. The four recommended uses of these grants would be: construction of a public safety network, including use of commercial infrastructure; coverage of rural areas; hardening existing commercial networks for public safety use, including reimbursement of non-recoverable engineering costs; and deployable capabilities for public safety.

The NBP provided an estimate of up to $6.5 billion for capital expenditures over ten years and operating costs of $1.3 billion a year. A subsequent report providing details on these projections were later released.[27] The report included a comparison of costs that concluded that building a dedicated public safety broadband network would require $15.7 billion in capital expenditures. A substantial part of the projected savings would come from the ability for public safety to use commercial towers.

The NBP stated that it was "essential that the United States establish a long-term, sustainable and adequate funding mechanism to help pay for the operation, maintenance and upgrade of the public safety broadband network."[28] To provide these funds, the plan recommended that a "minimal public safety fee" be assessed on all U.S. broadband users.

Legislative Proposals for Funding Infrastructure

The Broadband for First Responders Act of 2010 would rely on existing authorizations for funding public safety communications, such as the Interoperable Emergency Communications Grant program. This law authorized appropriations as necessary for FY2008 and appropriations of up to $400 million for fiscal years 2009 though 2012, with such sums as may be necessary in subsequent years.[29] The first appropriations were provided for FY2008 in the amount of $50 million, appropriations in subsequent years have also been for $50 million.[30]

The First Responders Protection Act of 2010, the Public Safety Spectrum and Wireless Innovation Act, and the discussion draft of the Public Safety Broadband Act of 2010 would fund network construction and operation with proceeds from future spectrum auctions. The bills have designated several sets of spectrum bands for auction as sources of revenue over a specified time period. The D Block is included in the draft discussion bill. Two funds would be created to receive auction proceeds. The first $5.5 billion would be destined for a Construction Fund; subsequent proceeds would be administered through a Maintenance and Operation Fund. The NTIA would have primary responsibility for grants programs covered by the funds. Construction projects that would be eligible would be for new construction for a public safety broadband network; improvements to existing commercial networks and other improvements to infrastructure needed to operate an interoperable, public safety broadband network in the 700 MHz band. The bills have described eligibility for reimbursement of maintenance and operational costs and related provisions.

The Next Generation Public Safety Device Act of 2010 would fund the $70 million authorized for RoIP development through revenues generated by future auctions. It would create a Public Safety Devices Communications Fund and authorize the NTIA to borrow the needed funds. The Senate version of the bill would place a ceiling on the amount that could be used to administer the program.

Spectrum Auctions as a Source of Funds

Congress has twice enacted laws to create special funds to hold the revenue of certain spectrum auctions for specific purposes. These funds represent a departure from existing practice, which requires that auction proceeds be credited directly to the Treasury as income. The Deficit Reduction Act of 2005 (P.L. 109-171, Title III) required the auctioning of licenses for spectrum currently used by TV broadcasters for analog transmissions. It established the Digital Television Transition and Public Safety Fund to receive this auction revenue and use some of the proceeds for the transition to digital television, public safety communications, and other programs. The Commercial Spectrum Enhancement Act (P.L. 108-494, Title II) established a Spectrum Relocation Fund to hold the proceeds of certain spectrum auctions for the specific purpose of reimbursing federal entities for the costs of moving to new frequency assignments.

The Public Safety Interoperable Communications (PSIC) grant program, now administered through the Department of Homeland Security, was funded under provisions in the Deficit Reduction Act of 2005.

CONCLUSION

Homeland security depends on effective communications for prevention, preparedness, and response to a range of threats. First responders and the larger public safety community that supports them rely heavily on effective radio communications to meet their responsibilities for homeland security. More important to the average American is the role that public safety services play in daily life and in responding to natural disasters. Flash floods, forest fires, tornados, hurricanes—Mother Nature provides endless variations for the scenarios of response and recovery.

The FCC and the Department of Homeland Security (DHS) have different perspectives on radio technology and infrastructure. DHS policies favor reliability and familiarity in their requirements and guidelines for technology and in their emphasis on training and repeated use of equipment. Spectrum policy at the FCC promotes spectrum efficiency and competition among commercial license-holders.

Congress has separately conferred authority on DHS and the FCC to act on behalf of public safety. In the case of DHS, this includes requirements to coordinate and support specific goals, such as interoperability and a national communications capability.[31] None of the actions required of DHS by Congress relate specifically to using 700 MHz spectrum to achieve these objectives. The FCC brings to the process several important mandates from Congress, such as an obligation to "promote safety of life and property through the use of wire and radio communication,"[32] as well as specific instructions regarding the assignment of frequencies at 700 MHz. In its National Broadband Plan, the FCC proposed that it assume the needed leadership role and has since taken a number of steps to realize the goals it has set for itself.

The bills introduced so far would increase the powers and responsibilities of the FCC in shaping the public safety communications network of the future, placing DHS in an advisory role. Governance of the public safety network at a national level would be dependent almost entirely on the FCC and its willingness to write and enforce regulations. The Public Safety Broadband Licensee Board of Governors and the Technical Advisory Committee of ERIC appear to be the primary conduits for presenting public safety views and requirements to the FCC. The Public Safety Broadband License, the charter for the Public Safety Broadband Licensee, and ERIC were established by the FCC through its rule-making process.

Since September 11, 2001, Congress has passed several laws that empowered the Department of Homeland Security to recognize and respond to technological developments in wireless and Internet Protocol (IP) communications, and to apply this knowledge to guiding the development of a nationwide, interoperable network for public safety. By choosing to focus on interim solutions, the Department seems to have passed on the opportunity to provide the needed leadership and planning to move public safety toward a next-generation communications network.

It would appear that neither agency has the needed depth of experience or resources to develop and deploy a leading-edge broadband network in a timely, cost-efficient manner. DHS has experience in guiding public safety agencies to meet Department-imposed requirements and guidelines. The FCC has experience in regulating spectrum use by both commercial and state and local (but not federal) public safety sectors. The FCC-created entities that might draw on a broader pool of skills and knowledge are dependent on the long-term, continuing, and unwavering interest of the FCC in maintaining and empowering them.

A well-grounded but flexible governance structure is critical to the future of public safety communications if it is to be national in scope, interoperable, and cost-effective. In particular, the cost of radios must be brought down to a competitive price range. The latest radios developed for public safety by DHS, the multi-band radios, are estimated to cost between $4,000 to $6,000.[33] The current narrowband radios being used for 700 MHz networks typically start at $3,000. Depending on the area of the country, the cost of ten radios would pay the salary of a teacher, or a teacher's aide. What might be the consequences of a federal policy that forces communities to decide between radios for their police force or teachers for their children, radios to fight fires or funds to maintain parks? Experience with early deployments of narrowband networks appears to indicate that many communities are buying two, four, maybe ten radios to provide for interoperable communications among senior level emergency managers, leaving the rest of their first responder resources to make do with existing equipment that might not work on the new network. At present, federal funds appear to go to build a sort of state-of-the-art wireless highway that only the radio equivalent of Hummers can use.

Several states and urban areas have submitted detailed plans to the FCC for building the nation's first broadband networks that might serve as a practical framework for evaluating policy options. These plans, developed according to FCC requirements, share many common features. Notable from a policy point of few are several recommendations that provide a common theme in these early submissions. These may be summarized as: (1) sufficient funding is essential; (2) networks that either cover an area designated as eligible for Urban Area Security Initiative programs, or cover a regional area—that is, large and/or densely populated areas—are more efficient to build, operate, and govern; (3) several critical technologies and standards, such as for radios, must be developed before the networks can be fully effective; (4) some form of governing sur-structure must be in place to assure uniformity of core operations while allowing for local customization of public safety applications; and (5) collaboration with commercial partners is important for mustering all the skills and knowledge resources needed for developing the leading-edge broadband networks that are the goals of the submitted plans.

A governance structure that can deliver these elements might meet the needs described by many representatives of the public safety community. It does not exist.

APPENDIX A. CONGRESSIONAL EFFORTS ON BEHALF OF PUBLIC SAFETY COMMUNICATIONS

Many of the statutes passed since 2001 have provided guidelines and set performance goals for public safety communications while delegating decisions about implementation to federal agencies and state officials. Although Congress has appropriated money for public safety communications it has not directly addressed the question of investment in network infrastructure, leaving it largely to federal agencies to set priorities for how public safety grants can be used. Most of the grant programs are now administered through the Department of Homeland Security (DHS).[34] Grants for emergency communications have been used to purchase equipment that facilitates interoperability, for planning, and for training.

To facilitate planning and coordination, and to provide direction, Congress authorized the creation of an Office of Emergency Communications (OEC) within DHS. The OEC was given the responsibility of preparing a National Emergency Communications Plan (NECP). The resulting plan set goals for improving emergency communications and interoperability but did not address developing a network infrastructure for public safety communications or for using the 700 MHz spectrum for that purpose.[35]

To support its vision[36] of interoperability as a system of systems, DHS sponsored an Emergency Response Council (ERC) composed of several dozen agencies, associations, and other entities involved in public safety and emergency response planning. In 2007 the ERC provided a set of agreements on a Nationwide Plan for Interoperable Communications. The ERC published 12 guiding principles deemed essential to their key goals of forging partnerships, designing interoperable systems, educating policymakers, and allocating resources.[37] To date, the council's role has been primarily to establish a base for advocacy and communication among representatives of public safety agencies and associations.

Congress first addressed the issue of emergency communications interoperability in the Homeland Security Act of 2002 (P.L. 107-296). Two years later, responding to recommendations of the National Commission on Terrorist Attacks Upon the United States (9/11 Commission), Congress included a section in the Intelligence Reform and Terrorism Prevention Act of 2004 (P.L. 108-458) that expanded its requirements for action in improving interoperability and public safety communications. Also in response to a recommendation by the 9/11 Commission,

Congress set a firm deadline for the release of radio frequency spectrum needed for public safety radios, as part of the Deficit Reduction Act of 2005 (P.L. 109-171). These laws provided the base from which the Department of Homeland Security (DHS) could develop a national public safety communications capability as required by the Homeland Security Appropriations Act, 2007 (P.L. 109-295). Title VI, Subtitle D of the act, referred to as the 21st Century Emergency Communications Act of 2006, placed new requirements on DHS. Additional requirements were included in the Implementing Recommendations of the 9/11 Commission Act of 2007 (P.L. 110- 53).

Balanced Budget Act of 1997

The initial allocation to public safety of frequencies in the 700 MHz band was required by Congress in the Balanced Budget Act of 1997 (P.L. 105-33),[38] which directed the Federal Communications Commission (FCC) to designate 24 MHz of spectrum capacity for public safety. To carry out the process of assigning this newly allocated spectrum asset, the FCC created the Public Safety National Coordination Committee (NCC) as a Federal Advisory Committee. Active from 1999 through 2003, the NCC had a Steering Committee from government, the public safety community, and the telecommunications industry. The NCC developed technical and operational recommendations for the 700 MHz band, including plans for interoperable channels. The existing governance for these channels is through Regional Planning Committees (RPCs),[39] established and loosely coordinated by the FCC, with the participation of the National Public Safety Telecommunications Council (NPSTC), a group consisting primarily of public safety associations. The RPCs are responsible for submitting 700 MHz band plans to the FCC for approval, and for managing these plans.

The Homeland Security Act of 2002 and Actions by the Department

Provisions of the Homeland Security Act instructed DHS to address some of the issues concerning public safety communications in emergency preparedness and response and in providing critical infrastructure. Telecommunications for first responders is mentioned in several sections, with specific emphasis on technology for interoperability.[40]

The newly created DHS placed responsibility for interoperable communications within the Directorate for Science and Technology, reasoning that the focus of DHS efforts would be on standards and on encouraging research and development for communications technology. Responsibility to coordinate and rationalize federal networks, and to support interoperability, had previously been assigned to the Wireless Public SAFEty Interoperable COMmunications Program—called Project SAFECOM—by the Office of Management and Budget as an e-government initiative. With the support of the George W. Bush Administration, SAFECOM was placed in the Science and Technology directorate and became the lead agency for coordinating federal programs for interoperability.[41] The Secretary of Homeland Security assigned the responsibility of preparing a national strategy for communications interoperability to the Office of Interoperability and Compatibility (OIC), which DHS created, an organizational move that was later ratified by Congress in the Intelligence Reform and Terrorism Prevention Act.[42] SAFECOM continued to operate as an entity within the OIC, which assumed the leadership role.

Intelligence Reform and Terrorism Prevention Act

Acting on recommendations made by the 9/11 Commission, Congress included several sections regarding improvements in communications capacity—including clarifications to the Homeland Security Act—in the Intelligence Reform and Terrorism Prevention Act (P.L. 108-458).

The Commission's analysis of communications difficulties on September 11, 2001, was summarized in the following recommendation.

> Congress should support pending legislation which provides for the expedited and increased assignment of radio spectrum for public safety purposes. Furthermore, high-risk urban areas such as New York City and Washington, D.C., should establish signal corps units to ensure communications connectivity between and among civilian authorities, local first

> responders, and the National Guard. Federal funding of such units should be given high priority by Congress.[43]

Congress addressed both the context and the specifics of the recommendation for signal corps capabilities. The Intelligence Reform and Terrorism Prevention Act amended the Homeland Security Act to specify that DHS give priority to the rapid establishment of interoperable capacity in urban and other areas determined to be at high risk from terrorist attack. The Secretary of Homeland Security was required to work with the Federal Communications Commission (FCC), the Secretary of Defense, and the appropriate state and local authorities to provide technical guidance, training, and other assistance as appropriate. Minimum capabilities were to be established for "all levels of government agencies," first responders, and others, including the ability to communicate with each other.[44] The act further required the Secretary of Homeland Security to establish at least two trial programs in high-threat areas. The process of development for these programs was to contribute to the creation and implementation of a national model strategic plan. The purpose was to foster interagency communications at all levels of the response effort. Building on the concept of using the Army Signal Corps as a model, the law directed the Secretary to consult with the Secretary of Defense in the development of the test projects, including review of standards, equipment, and protocols.[45]

Congress also raised the bar for performance and accountability, setting program goals for the Department of Homeland Security. Briefly, the goals were to:

- Establish a comprehensive, national approach for achieving interoperability;
- Coordinate with other federal agencies;
- Develop appropriate minimum capabilities for interoperability;
- Accelerate development of voluntary standards;
- Encourage open architecture and commercial products;
- Assist other agencies with research and development;
- Prioritize, within DHS, research, development, testing and related programs;
- Establish coordinated guidance for federal grant programs;
- Provide technical assistance; and
- Develop and disseminate best practices.

The act included a requirement that any request for funding from DHS for interoperable communications "for emergency response providers" be accompanied by an Interoperable Communications Plan, approved by the Secretary. Criteria for the plan were also provided in the act.[46]

The act also provided a sense of Congress that the next Congress—the 109th—should pass legislation supporting the Commission's recommendation to expedite the release of spectrum. This was addressed in the Deficit Reduction Act of 2005 (P.L. 109-171).

The Homeland Security Appropriations Act, 2007

The destruction caused by Hurricanes Katrina and Rita in August-September 2005 reinforced the recognition of the need for providing interoperable, interchangeable communications systems for public safety and also revealed the potential weaknesses in existing systems to withstand or recover from catastrophic events. Testimony at numerous hearings following the hurricanes suggested that DHS was responding minimally to congressional mandates for action, most notably as expressed in the language of the Intelligence Reform and Terrorism Prevention Act. Bills subsequently introduced in both the House and the Senate proposed strengthening emergency communications leadership and expanding the scope of the efforts for improvement. Some of these proposals were included in Title VI of the Homeland Security Appropriations Act, 2007 (P.L. 109-295). Title VI—the Post-Katrina Emergency Management Reform Act of 2006— reorganized the Federal Emergency Management Agency (FEMA), gave the agency new powers, and clarified its functions and authorities within DHS.[47]

The act also addressed public safety communications in Title VI, Subtitle D—the 21st Century Emergency Communications Act of 2006. This section created an Office of Emergency Communications (OEC)) and the position of Director, reporting to the Assistant Secretary for Cybersecurity and Communications. The Director was required to take numerous steps to coordinate emergency communications planning, preparedness, and response, particularly at the state and regional level. These efforts were to include coordination with Regional Administrators appointed by the FEMA Administrator to head ten Regional Offices. To assist these efforts, Congress required the creation of Regional Emergency Communications Coordination (RECC) Working Groups.[48]

Other responsibilities assigned to the Director included conducting outreach programs, providing technical assistance, coordinating regional working groups, promoting the development of standard operating procedures and best practices, establishing non-proprietary standards for interoperability, developing a national communications plan, working to assure operability and interoperability of communications systems for emergency response, and reviewing grants. Required elements of the National Emergency Communications Plan included establishing requirements for assessments and reports, and an evaluation of the feasibility of developing a mobile communications capability modeled on the Army Signal Corps. The feasibility study was to be done by DHS on its own or in cooperation with the Department of Defense. Congress also required assessments of emergency communications capabilities, including an inventory that identified radio frequencies used by federal departments and agencies.[49]

Many of the functions Congress envisioned for the OEC were later assumed by the Command, Control and Interoperability Division in the Directorate of Science and Technology.

Regional Emergency Communication Coordination

Congress directed the OEC to coordinate with the Regional Emergency Communication Coordination (RECC) Working Groups established by FEMA.[50] These groups could provide a platform for coordinating emergency communications plans among states and were intended to include representatives from many sectors with responsibility for public safety and security. Plans for forming RECCs were announced in December 2007. In 2008 organization charts were developed, graphing how the RECCs were structured and where they would fit in the existing chain-of-command of the Federal Emergency Management Agency (FEMA). A National RECC Coordinator was appointed and plans were announced to appoint administrators for each of the regions.

A key proposal for the RECC structure is to "Establish and use the RECC's as a single Federal emergency communications coordination point for Federal interaction with the State, local and tribal governments."[51] It is not clear at this early stage whether the RECCs will become an effective conduit for interaction to develop policies and plan for shared infrastructure or a forum for FEMA's Disaster Operations Directorate to relay guidelines and orders. Congress placed an emphasis on assisting first responders in its statement of

RECC goals but did not limit the RECCs' ability to set more inclusive goals. Based on the role of RECCs as assigned by the National Emergency Communications Plan, their focus will be narrowly on assisting first responders to prepare for disaster response. Leadership will be provided by FEMA and governance will be through the chain-of-command of the agencies' directorates.

The formation of the regional working groups, the RECCs, responded in part to requests from the public safety community to expand interoperable communications planning to include the second tier of emergency workers. Non-federal members of the RECC are to include first responders, state and local officials and emergency managers, and public safety answering points (911 call centers). Additionally, RECC working groups are to coordinate with a variety of communications providers (such as wireless carriers and cable operators), hospitals, utilities, emergency evacuation transit services, ambulance services, amateur radio operators, and others as appropriate.

National Emergency Communications Plan

In compliance with requirements of the Homeland Security Appropriations Act, 2007, the Department of Homeland Security issued the National Emergency Communications Plan (NECP) in July 2008.[52]

The NECP sets three goals for levels of interoperability[53]

- By 2010, 90% of all areas designated within the Urban Areas Security Initiative (UASI) will demonstrate response-level emergency communications, as defined in grant programs, within one hour for routine events involving multiple jurisdictions and agencies.
- By 2011, 75% of non-UASI will have achieved the goal set for UASIs.
- By 2013, 75% of all jurisdictions will be able to demonstrate response-level emergency communications within three hours for a significant incident as outlined in national planning scenarios.

These jurisdictional goals are to be knit together into a national communications capability through program efforts such as FEMA's Regional Emergency Communications Coordination (RECC) Working Group. The three goals are bolstered by seven objectives for improving emergency communications for first responders, dealing largely with organization and coordination.[54] Each of these objectives have "Supporting Initiatives" and milestones.

Deficit Reduction Act of 2005 and the Public Safety Interoperability Grant Program

Provisions in the Deficit Reduction Act of 2005 planned for the release of spectrum by February 18, 2009[55] and created a fund to receive spectrum auction proceeds and disburse designated sums to the Treasury and for other purposes,[56] including a grant program of up to $1 billion for public safety agencies. The fund's disbursements were to be administered by the NTIA.[57] At the time, the Congressional Budget Office projected that the grants program for public safety would receive $100 million in FY2007, $370 million in FY2008, $310 million in FY2009 and $220 million in FY2010.[58] However, the 109th Congress, in its closing hours, passed a bill with a provision requiring that the grants program receive "no less than" $1 billion to be awarded "no later than" September 30, 2007.[59] Language in Implementing Recommendations of the 9/11 Commission Act of 2007 (P.L. 110-53) required some changes in the grant program and reaffirmed the 2007 fiscal year deadline.[60]

In February 2007, the NTIA transferred the management of the public safety grant program to DHS, signing a memorandum of understanding (MOU) with the Office of Grants and Training.[61] The MOU included an overview of how the Public Safety Interoperable Communications (PSIC) Grant Program, as it is called, is to be administered. The overview was reiterated and explained in testimony.[62] Both the MOU and the testimony indicate that the priority was to fund needs identified through Tactical Interoperable Communications Plans and Statewide Interoperable Plans developed in conjunction with SAFECOM.

On July 18, 2007, the Secretaries of Commerce and Homeland Security jointly announced the details of the PSIC grant program.[63] The program, as announced, was to provide $968,385,000 in funding for all 50 states, the District of Columbia, and U.S. Territories.[64] The announcement of the top-level, statewide allocations met the September 30 deadline set by Congress. The states, however, have additional time to submit their detailed requests. Originally, states were eligible to receive funds through FY2010.[65] New legislation (P.L. 111-96) extends the deadline through FY2012.[66] The status of the PSIC grant program was discussed at a hearing in March 2009. Testimony at the time indicated that all of the states, territories, and the District of

Columbia had filed Statewide Communication Interoperability Plans, a prerequisite for receiving funds.[67]

APPENDIX B. SPECTRUM CHART

Below is an excerpt of the 700 MHz band plan that shows the location of public safety allocations and the D Block, and their relation to other adjacent spectrum holdings.

The figure below represents the relative locations on the spectrum chart of frequencies assigned for public safety and commercial purposes. The accompanying legend provides the key to the type of license and the amount of spectrum associated with that spectrum. For example, reading from the left, the first band represents 11 MHz of the C Block, which is for commercial purposes. The 11-MHz allocation is in two assignments, at 746-757 MHz and at 776-787MHz. The controversial D Block is two assignments of 5 MHz each; these are contiguous with the two assignments for the Public Safety Broadband License.

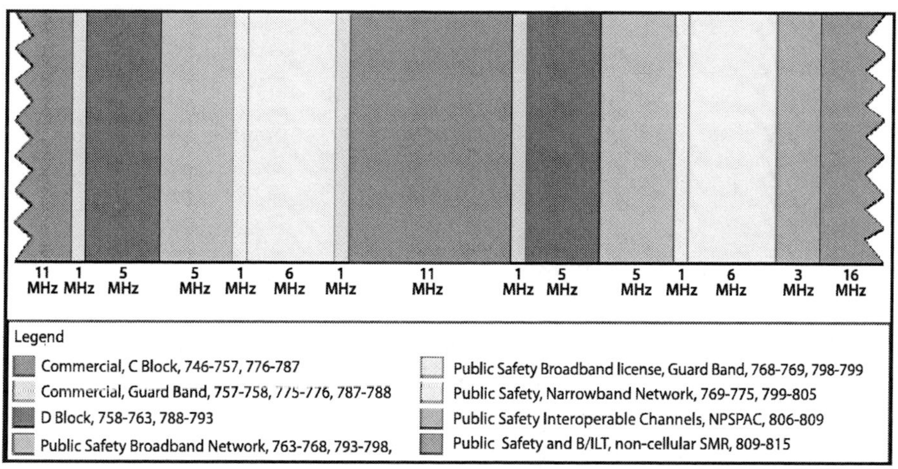

Source: Federal Communications Commission, Revised 700 MHz Band Plan for Commercial Services.

Figure B-1. Public Safety and the D Block.

Note that the figure represents only part of the entire 700 MHz band.

Appendix C. Managing Technology and Spectrum Resources

Within the federal government, the Department of Homeland Security (DHS) has authority for planning and implementing public safety communications solutions. The Federal Communications Commission (FCC) created a Public Safety and Homeland Security Bureau in 2006 to consolidate its many programs oriented toward public safety.[68] The FCC and DHS have each approached the goal of communications interoperability from a different perspective.

The following discussion provides snapshot summaries of the approaches adopted by the two agencies and the technologies and network design concepts that might be applied by the FCC.

The section also discusses the trend to Internet-based concepts for networks and spectrum management.

Ideas for managing emergency communications have moved along an evolutionary path from the 1990s, when agreement was reached on developing standards for interoperable communications, to the system-of-systems concept embraced by DHS, to the network-oriented proposals of the FCC, Public Safety Spectrum Trust (PSST), National Public Safety Telecommunications Council (NPSTC),[69] and others. The ubiquity of the Internet and the standards that support it are leading to a new path for managing spectrum and network resources.

System-of-Systems

The communications solutions advocated by DHS have focused on developing what is often referred to as a "system of systems." The choice of terminology implies that independent systems are made to work with each other through bridges and gateways that connect incompatible technology choices into a larger system.

This approach maximizes the value of past investments but does not represent an efficient use of resources. Backward-compatible radio equipment that can support several generations of different technologies, for example, is more expensive than equipment designed to work only with newer network technology. Spectrum usage is inefficient because more than one channel is often used to convey a single communication from system to system.

Essentially, the system-of-systems concept starts with the radio user and works its way up, adding and connecting the different levels of command and control needed to respond to specific situations. DHS refers to this as a practitioner-driven approach.

Many of the DHS programs for public safety have focused on achieving interoperability within the existing framework of proprietary systems and by expanding the diffusion of Project 25, or P25, standards.[70] Backward compatibility with legacy systems is one of the principles behind the digital radio and interoperable gateway[71] standards of P25. Its use is advocated by many public safety agencies and by DHS.[72] P25-compliant technologies coordinate and connect specified radio channels.[73]

Currently, the Command, Control, and Interoperability Division of the Science and Technology Directorate at DHS is testing and evaluating P25 multi-band radios.[74]

The initial phase of the program was announced July 1, 2009. Results will be documented in a report that "will provide details to manufacturers about the needs of the response community and assist officials in making informed radio purchasing decisions in the future."[75]

Shared Networks

The FCC has taken a more network-oriented approach to achieving interoperability by laying out a plan for a national network at 700 MHz that would eventually reach every community with the same technology and connectivity, providing a common base for individual applications. Network-centric solutions start with the network framework, which sets a common standard.

Any traffic that wants to use this network has to accommodate that standard (although it can use additional standards as well). Network-centric solutions tend to be managed from the top down, with centralized control of core decisions.

The FCC, primarily through the Emergency Response Interoperability Center (ERIC), is attempting to structure a central administration to lead the decision-making process for implementing a nationwide public safety network.

Interconnected Networks

The PSST was assigned the Public Safety Broadband License (PSBL) as part of the FCC's plans to create a public-private partnership. The PSST considers that the new broadband network will serve primarily as a data exchange network (text, photos, video, etc.) that would operate as an adjunct to the current mission critical public safety voice systems. Existing voice communications systems and new narrowband systems at 700 MHz would operate independently of the broadband network with an interface to be established in a future development phase.[76]

The Board of Directors of the PSST received recommendations from NPSTC on how to achieve interoperability among different public safety networks operating in 700 MHz allocations for public safety broadband. These were submitted to the FCC for consideration in December 2009.[77] The recommendations were oriented toward paving the way for the early construction of networks by states and cities. The interoperable framework provided by the task force is based on connecting independent public safety networks. Interoperability would be facilitated by a number of guiding principles and requirements, such as access to the Internet and IP-based voice interoperability gateways. In general, the recommendations of the task force would facilitate these expectations.

- Regional (including state and local) broadband systems will operate within the framework of a Nationwide Broadband Data System (NBDS).
- The NBDS will use Long Term Evolution (LTE) technology and it is assumed that regional systems will as well.
- Defining minimum requirements for public safety broadband networks at 700 MHz will enable national interoperability.
- An advisory council will provide governance among individual operators and the PSBL.
- Public-private partnerships will be allowed.
- Different scenarios for assignment of the D Block will be accommodated by the Task Force requirements and recommendations.
- Regional operators will have the right to deploy systems in advance of final requirements and to select and deploy applications beyond what is required.

- Technical requirements will be specified to facilitate roaming and interoperability.
- Those that have filed requests to the FCC for permission to build systems will be able to fulfill their 700 MHz broadband objectives as quickly as possible.
- Best practices for network architecture and configurations will be provided but not required.

Among other recommendations made to the FCC in the same filing, the PSST asked the FCC to authorize it "to establish the technology standard for the 700 MHz nationwide public safety broadband network...."

The Association for Public-Safety Communications Officials—International (APCO) has announced its intention to develop standards for the broadband networks at 700 MHz. Specifically, APCO "will identify gaps and set standards in those areas where none currently exist and where standards are necessary to ensure roaming and interoperability ..." and will "establish basic requirements necessary to ensure interoperability" for the network.[78] APCO is accredited by the American National Standards Institute as a Standards Development Organization.

IP-Enabled Networks

As part of the discussion about how to bring broadband to public safety users, several organizations recommended Long Term Evolution (LTE), a fourth-generation wireless technology,[79] for the underlying network infrastructure on the 700 MHz frequencies.[80] The FCC has concluded that it will require LTE technology for the network infrastructure for the D Block and the PSBL.[81] Fourth-generation technologies such as LTE are being developed to use Internet Protocol (IP) standards, assuring a high degree of interoperability among other IP-based technologies. Developing standards for public safety interfaces on LTE networks could represent a shift in concept for public safety communications—to IP-based platforms and communications management at the network level. Some public safety representatives have shown a willingness to move from a model that connects disparate systems to a model that provides interoperability through network administration.[82]

Some states have decided to deploy IP-enabled fiber optic networks to support their communications needs, including those of public safety. These networks use IP standards to achieve the same level of interoperability, availability, and flexibility associated with the Internet but do not necessarily link to the Internet.

Congress has recognized the value of IP-based networks for 911 communications by, for example, requiring the NTIA and the National Highway Traffic Safety Administration to prepare recommendations that would support the transition of out-dated 911 systems to IP-based technologies.[83] Congress has not previously considered giving the same attention to the adoption of IP-based technologies for public safety radio communications.

Adaptive Network Technologies

The FCC, DHS, PSST, and NPSTC approaches to interoperability, although different in perspective, are all based on managing radio channels as the way to meet common goals such as minimizing interference among wireless transmissions.

The concept of channel management dates to the development of the radio telegraph by Guglielmo Marconi and his contemporaries. In the United States, mitigation of radio interference was addressed in what is commonly known as the Radio Act of 1912. Passage of the bill, versions of which had been introduced in earlier Congresses, was prompted in part by Marconi's testimony at a congressional hearing investigating the sinking of the Titanic. The act established the basic principle of assigning licenses for specific channels through a central federal authority, which became the FCC with the passage of the Communications Act of 1934.

In the age of the Internet, however, channel management is an inefficient way to provide spectrum capacity for mobile broadband.[84] Innovation points to network-centric spectrum management as an effective way to provide spectrum capacity to meet the bandwidth needs of fourth-generation wireless devices. Network-centric technologies organize the transmission of radio signals along the same principle as the Internet. A transmission moves from origination to destination not along a fixed path but by passing from one available node to the next. Pooling resources, one of the concepts that powers the Internet now, is likely to become the dominant principle for spectrum management in the future.

The new generations of iPhones and Android-based[85] mobile devices provide early examples of how the Internet is likely to change wireless communications as more and more of the underlying network infrastructure is converted to IP-based standards. The devices use Internet protocols to perform many of its functions; these require time and space—spectrum capacity—to operate.

The core Internet Protocol (TCP/IP) was conceived to work with high capacity landline networks. In a wireless environment, IP applications are bandwidth-intensive, consuming large amounts of channel capacity. Although future generations of mobile broadband devices will no doubt use IP applications that have been refined for the wireless environment, additional capacity will still be required to handle expected increases in activity.

More efficient spectrum use can be realized by integrating adaptive networking technologies, such as dynamic spectrum access (DSA),[86] with IP-based commercial network technologies such as LTE. Radios using DSA chipsets are more effective at managing interference and congestion than the channel management techniques currently in use. If a channel's link fails, the radio is cut off. When radios are networked using DSA, individual communications nodes continue to operate and can compensate for failed links. The effects of interference are manageable rather than catastrophic. The network is used to overcome radio limitations.

Adaptive networking has the potential to organize radio communications to achieve the same kinds of benefits that have been seen to accrue with the transition from proprietary data networks to the Internet. Adaptive technologies are designed to use pooled spectrum resources. Pooling spectrum licenses goes beyond sharing. Licenses are aggregated and specific ownership of channels becomes secondary to the common goal of maximizing network performance. For many, the construction of a new network for public safety communications represents an opportunity to reap the perceived benefits of shared infrastructure and pooled spectrum by using the technologies and principles of network-centric operations.

End Notes

[1] H.R. 5081, the Broadband for First Responders Act of 2010, Representative King, introduced April 20, 2010, referred to the House Committee on Energy and Commerce and S. 3625, the First Responders Protection Act of 2010, Senator Lieberman, and S. 3756, Public Safety Spectrum and Wireless Innovation Act, Senator Rockefeller, both referred to the Senate Committee on Commerce, Science, and Transportation.

[2] Staff discussion draft, the Public Safety Broadband Act of 2010, House Committee on Energy and Commerce, Subcommittee on Communications, Technology, and the Internet,

presented June 14, 2010; http://energycommerce.house.gov/documents/2 0100615 /Public. Safety.Broadband.Act.Discussion.Draft.pdf.

[3] Spectrum is measured in cycles per second, or hertz. Standard abbreviations for measuring frequencies include kHz—kilohertz or thousands of hertz; MHz—megahertz, or millions of hertz; and GHz—gigahertz, or billions of hertz.

[4] The Deficit Reduction Act of 2005, P.L. 109-171, Title III, Sec. 3002 120 STAT. 21 set a deadline for releasing the frequencies. Initial legislation requiring the release was in the Balanced Budget Act of 1997, P.L. 105-33, Title III, Sec. 3003 and Sec. 3004, 111 STAT. 265 et seq.

[5] Spectrum resources are typically segmented into bands of radio frequencies. The 700 MHz band includes radio frequencies from 698 MHz to 806 MHz. Public safety has frequency allocations within this band totaling 24 MHz.

[6] Expediting the release of these frequencies was among the recommendations of the 9/11 Commission. The 9/11 Commission Report: Final Report of the National Commission on Terrorist Attacks Upon the United States, p. 397, Washington: GPO, 2004.

[7] The Deficit Reduction Act of 2005, P.L. 109-171, Title III, Sec. 3003 120 STAT. 22.

[8] A summary of FCC actions regarding the D Block is included as Background in FCC, Order, released May 12, 2010, PS Docket No. 06-229, concerning waivers to allow early establishment of public safety broadband networks at 700 MHz, at http://fjallfoss.fcc. gov/edocs_public/attachmatch/FCC-10-79A1.pdf.

[9] FCC, Connecting America: The National Broadband Plan, http://www.broadband.gov/ download-plan/.

[10] Connecting America, Recommendation 5.8.2.

[11] Connecting America, Recommendation 16.1. FCC Order establishing ERIC was released April 23, 2010, http://fjallfoss.fcc.gov/edocs_public/attachmatch/FCC-10-67A1.pdf.

[12] These requirements are included in presentations by Ralph A. Haller, Chairman of the National Public Safety Telecommunications Council, and Chief Harlin R. McEwen, Chairman of the Public Safety Spectrum Trust, at an FCC National Broadband Plan Staff Workshop on August 25, 2009. The presentations are available at http://www.npstc. org/ index.jsp.

[13] Broadband refers to the capacity of the radio frequency channel. A broadband channel can transmit live video, complex graphics and other data-rich information as well as voice and text messages whereas a narrowband channel might be limited to handling voice, text, and some graphics.

[14] One frequently cited definition of interoperability has been provided by the government agency SAFECOM: "In general, interoperability refers to the ability of public safety emergency responders to work seamlessly with other systems or products without any special effort. Wireless communications interoperability specifically refers to the ability of public safety officials to share information via voice and data signals on demand, in real time, when needed, and as authorized." http://www.safecomprogram.gov.

[15] The practice of transferring a wireless call from one network to another—or roaming—is described in Understanding Wireless Telephone Coverage Areas, FCC Consumer Facts at http://www.ifap.ru/library/book385.pdf.

[16] Connecting America, Recommendation 16.1.

[17] FCC, Order, PS Docket No. 06-229, released April 23, 2010 at http://fjallfoss.fcc.gov/ edocs_public/attachmatch/FCC-10-67A1.pdf.

[18] NIST, "Demonstration Network Planned for Public Safety 700 MHz Broadband," December 15, 2009 at http://www.nist.gov/eeel/oles/network_121509.cfm.

[19] FCC, Order, Request for Waiver of Various Petitioners to Allow the Establishment of 700 MHz Interoperable Public Safety Wireless Broadband Networks, P.S. Docket No. 06-229, released May 12, 2010 at http://fjallfoss.fcc.gov/edocs_public/attachmatch/FCC-10-79A1.pdf.

[20] These are: Adams County, CO; the State of Alabama; Boston, MA; Calumet, Outagamie, and Winnebago Counties, WI; Charlotte, NC; Chesapeake, VA;. the District of Columbia; the State of Mississippi; the State of New Mexico; New York City; the State of Oregon; Pembroke Pines, FL; San Antonio, TX; the San Francisco Bay Urban Area Security Initiative; and Seatttle, WA.

[21] FCC, Order, P.S. Docket No. 06-229, released August 17, 2010 at http://www.fcc.gov/Daily_Releases/Daily_Business/2010/db0817/DA-10-1540A1.pdf.

[22] FCC, Public Notice, "Public Safety and Homeland Security Bureau Offers Further Guidance to Conditional Waiver Recipients on Completing the Interoperability Showing Required by the 700 MHz Waiver Order," P.S. Docket No. 06- 227, DA 10-923, released May 21, 2010, at http://fjallfoss.fcc.gov/edocs_public/attachmatch/DA-10-923A1.pdf.

[23] 3rd Generation Partnership Project (3GPP) coordinate telecommunications standards bodies as "Organization Partners," see http://www.3gpp.org/About-3GPP. 3GPP is addressing commercial standards for 4th Generation technologies, including LTE; see http://www.3gpp.org/technologies.

[24] Dated July 20, 2010, at http://energycommerce.house.gov/documents/ 20100726/ Letter. FCC. 07.26.2010.pdf.

[25] Cyren Call Communications Corporation, in ex parte comments filed with the FCC on June 4, 2007, set the cumulative capital expenditure for building a public-private network at $18 billion, of which roughly a third of the cost would be for enhancements for public safety use. An estimate from Northrop-Grumman Corporation placed the cost at $30 billion, when service applications are included. (Statement by Mark S. Adams, Chief Architect Networks and Communications, at WCA 2007, Washington, DC, June 14, 2007.) These estimates do not include the cost of radios. An estimated range of $20 billion to $40 billion for network infrastructure was discussed at a House of Representatives hearing held by the Committee on Energy and Commerce, Subcommittee on Communications, Technology, and the Internet, "A National, Interoperable Broadband Network for Public Safety: Recent Developments," September 24, 2009.

[26] Connecting America, p. 317.

[27] FCC, A Broadband Network Cost Model: A Basis for Public Funding Essential to Bringing Nationwide Interoperable Communications to America's First Responders, OBI Technical Paper No. 2, May 2010 at http://www.fcc.gov/pshs/docs/ps-bb-cost-model.pdf.

[28] Connecting America, p. 319.

[29] P.L. 110-53, Title III, 121 STAT. 299; 6 U.S.C. 579.

[30] Background information on PSIC grants is provided in http://www.ntia.doc.gov/psic/ index.html.

[31] Discussed in detail in Appendix A, "Congressional Efforts on Behalf of Public Safety Communications."

[32] 47 U.S.C. § 151. The FCC relied partly on this authority in requiring the relocation of commercial licenses because transmissions were interfering with public safety radio communications in the 800 MHz band. See CRS Report RL32408, Spectrum Policy: Public Safety and Wireless Communications Interference, by Linda K. Moore.

[33] Department of Homeland Security, S&T Snapshots, - Command, Control, and Interoperability, "The Beginning of the End of the Single-Band Radio for Public Safety," October 20, 2009 at http://www.dhs.gov/files/programs/gc_1258141690101.shtm.

[34] DHS grants programs are discussed in CRS Reports CRS Report R40632, FY2010 Department of Homeland Security Assistance to States and Localities, and CRS Report R40246, Department of Homeland Security Assistance to States and Localities: A Summary and Issues for the 111th Congress, both by Shawn Reese.

[35] DHS, National Emergency Communications Plan, July 2008 at http://www.dhs.gov/xlibrary /assets/national_emergency_communications_plan.pdf

[36] "Our vision was developed at the 2003 SAFECOM/AGILE Joint Program Planning Meeting in San Diego, CA.", Emergency Response Council, Agreements on a Nationwide Plan for Interoperable Communications, Summer 2007, footnote 1.

[37] Op. cit., Agreements on a Nationwide Plan for Interoperable Communications.

[38] 47 U.S.C. § 309 (j) (14).

[39] Additional information at http://www.fcc.gov/pshs/public-safety-spectrum/700-MHz/regional-planning.html.

[40] Notably, P.L. 107-296, Sec. 232, 116 STAT. 2159 and Sec. 502, 116 STAT. 2213.

[41] "Homeland Security Starting Over With SAFECOM," Government Computer News, June 9, 2003.

[42] P.L. 108-458, Title VII, Subtitle C, Sec. 7303 (a) (2), 118 STAT. 3843-3844.

[43] The 9/11 Commission Report: Final Report of the National Commission on Terrorist Attacks Upon the United States, Washington: GPO, 2004, p. 397.

[44] P.L. 108-458, Title VII, Subtitle C, Sec. 7303, 118 STAT. 3843 et seq.

[45] P.L. 108-458, Title VII, Subtitle C, Sec. 7304, 118 STAT. 3847-3848.

[46] P.L. 108-458, Title VII, Subtitle C, Sec. 7303 118 STAT. 3843 et seq.

[47] Information on the FEMA reorganization is provided in CRS Report RL33729, Federal Emergency Management Policy Changes After Hurricane Katrina: A Summary of Statutory Provisions, coordinated by Keith Bea.

[48] P.L. 109-295, Title VI, Sec. 671(b), 'Title XVIII, 'Sec. 1805, 120 STAT. 1439.

[49] P.L. 109-295, Title VI, Sec. 671(b), 'Title XVIII, 'Sec. 1803, 120 STAT. 1437-1438.

[50] P.L. 109-295, Title VI, Sec. 671(b), "Title XVIII, "Sec. 1805, 120 STAT. 1439.

[51] Presentation by Brian Carney, National RECC Coordinator, Disaster Operations Directorate, Federal Emergency Management Agency, Department of Homeland Security, "Regional Emergency Communications Coordination Working Groups (RECCWG)" at National Public Safety Telecommunications Council (NPSTC), Seattle, Washington, September 15, 2008 at http://www.npstc.org/meetings/GB_Carney_ RECC%20Briefing_090208_NPSTC. ppt.

[52] DHS, National Emergency Communications Plan, July 2008 at http://www.dhs.gov/xlibrary /assets/national_emergency_communications_plan.pdf.

[53] National Emergency Communications Plan, "Executive Summary," page ES-1.

[54] Ibid., "Executive Summary," page ES-2.

[55] P.L. 109-171, Sec. 3002 (a) (1) (B). The deadline was later extended through June 12, 2009, by the DTV Delay Act,P.L. 111-4, Sec. 2 (a) (1).

[56] P.L. 109-171, Sec. 3004, 120 STAT. 22-23.

[57] P.L. 109-171, Sec. 3006, 120 STAT. 24-25.

[58] Congressional Budget Office Cost Estimate, S. 1932, Deficit Reduction Act of 2005, January 27, 2006, p. 21, http://www.cbo.gov/showdoc.cfm?index=7028&sequence=0.

[59] P.L. 109-459, Sec. 2.

[60] P.L. 110-53, Implementing Recommendations of the 9/11 Commission Act of 2007, Title XXII, Sec. 2201, 121 STAT. 537 et seq.

[61] MOU at http://www.ntia.doc.gov/otiahome/psic/PSICMOU_Executed_2-16-2007.pdf.

[62] Testimony of Corey Gruber, Acting Assistant Secretary for Grants and Planning, Office of Grants and Training, Department of Homeland Security at hearing on "Public Safety Interoperable Communications Grants: Are the Departments of Homeland Security and Commerce Effectively Coordinating to Meet our Nation's Emergency Communications Needs?" House of Representatives, Homeland Security Committee, Subcommittee on Emergency Communications, Preparedness, and Response, March 14, 2007.

[63] Press releases at http://www.dhs.gov/xnews/releases/pr_1184783934669.shtm and http://www.ntia.doc.gov/ntiahome/press/2007/PSIC_071807.pdf.

[64] See http://www.dhs.gov/xgovt/grants/gc_1184774852768.shtm. The NTIA website main page has a section devoted to PSIC at http://www.ntia.doc.gov.

[65] For details, see http://www.ntia.doc.gov/psic/awards.html.

[66] S. 1694, signed into law November 6, 2009.

[67] Testimony of W. Ross Ashley, III, Assistant Administrator, Grant Programs Directorate, FEMA, Hearing, "Interoperable Communications," House of Representatives, Committee on Appropriations, Subcommittee on Homeland Security, March 17, 2009.

[68] See http://www.fcc.gov/pshs/about-us/.

[69] See, for example, NPSTC, 700 MHz Public Safety Broadband Task Force Report and Recommendations, September 14, 2009 at http://www.npstc.org/documents/ 700_MHz_BBTF_Final_Report_0090904_v1_1.pdf.

[70] P25 conforms to recommendations made in 1996 by the Public Safety Wireless Advisory Committee (PSWAC) regarding the improvement of public safety communications over wireless networks, see "Final Report of the Public Safety Wireless Advisory Committee," September 11, 1996. The committee was disbanded after publication of its recommendations. The Association for Public-Safety Communications Officials—International (APCO) is a principle player in the development of P25 standards. Currently, meetings to develop standards are managed by the Telecommunications Industry Association, an ANSI-standards-setting body. See http://www.tiaonline.org/standards/technology/project_25/index.cfm/.

[71] Gateways, the current solution for interoperability, can connect radios using different technologies and frequencies.

[72] Guidance for standards are included, for example in "Recommended Guidance for Federal Grant Programs," at http://www.safecomprogram.gov/NR/rdonlyres/9DFFF882-1895-47F5-B724-9808BF1F9FE9/0/FY2009SAFECOM RecommendedGuidancefor FederalGrant ProgramsFINAL110408.pdf.

[73] A useful description of the development, benefits, and disadvantages of P25 was issued by the Department of Justice in 2007: Issue Brief Number 6, "Project 25: The Quest for Interoperable Radios," by Dan Hawkins, May 2007 at http://www.cops.usdoj.gov /files/ric/Publications/project25interopradios.pdf.

[74] Multi-band radios facilitate switching from one frequency to another in as many as four different bands.

[75] DHS Press Release, "DHS Announces Sites for Multi-Band Radio Pilot," July 1, 2009 at http://www.dhs.gov/ynews/releases/pr_1246478388904.shtm, and S&T Snapshots—Command, Control & Interoperability, "The Beginning of the End of the Single Band Radios for Public Safety," October 20, 2009, at http://www.dhs.gov/files/programs/ gc_1258141690101.shtm.

[76] This is the description supplied by PSST, the current Public Safety Broadband Licensee, in an e-mail from Chief Harlin R. McEwen to CRS on August 3, 2009. A similar description was provided in testimony by Chief McEwen at the hearing held by the House of Representatives, Committee on Energy and Commerce, Subcommittee on Communications, Technology, and the Internet: "A National, Interoperable Broadband Network for Public Safety: Recent Developments," September 24, 2009.

[77] Submitted December 15, 2009, as an Ex Parte communication, PS Docket No.06-229, WT Docket No. 06-150 at http://www.psst.org/documents/PSST-Minimum-Recommendations-121509.pdf.

[78] APCO News, "APCO Announces Intent to Develop Standards for Build Out of 700 MHz Broadband Network," October 14, 2009 at http://www.apcointl.org/new/news/ standards_700mhz.php.

[79] A discussion of broadband technology is include in CRS Report R40674, Spectrum Policy in the Age of Broadband: Issues for Congress , by Linda K. Moore.

[80] APCO and NENA jointly endorsed LTE . NPSTC, and the PSST are among those that also have endorsed LTE.

[81] This decision is discussed in FCC White Paper, The Public Safety Nationwide Interoperable Broadband Network: A New Model for Capacity, Performance and Cost, June 2010, at http://fcc.gov/pshs/docs/releases/DOC-298799A1.pdf.

[82] In an interview, Richard Mignon, the incoming president of APCO, observed that broadband is "the future of public safety communications.... It's almost like reinventing public safety technology and how we work together." As reported in MissionCritical Communications TRANSMISSION, e-newsletter, August 19, 2009.

[83] Next Generation 911 technologies and congressional actions in support of 911are discussed in CRS Report R41208, Emergency Communications: Broadband and the Future of 911 , by Linda K. Moore.

[84] A leading advocate for replacing channel management of radio frequency with network-centric management is Preston Marshall, the source for much of the information about network-centric technologies in this report. Mr. Marshall is Director, Information Sciences Institute, University of Southern California, Viterbi School of Engineering, Arlington, Virginia. CRS also spoke with other experts who provided background on the topic.

[85] Google, which uses the Android platform, describes it as "a software stack for mobile devices that includes an operating system, middleware, and key applications."

[86] Dynamic Spectrum Access, Content-Based Networking, and Delay and Disruption Technology Networking, along with cognitive radio, and decision-making software, are examples of technologies that can enable Internet-like management of spectrum resources.

In: Emergency Communications ISBN: 978-1- 62081-140-5
Editors: D. H. Martinez and A. E. Young © 2012 Nova Science Publishers, Inc.

Chapter 3

A Broadband Network Cost Model: A Basis for Public Funding Essential to Bringing Nationwide Interoperable Communications to America's First Responders OBI Technical Paper No. 2*

Federal Communications Commission

Executive Summary

In March 2010, the FCC released its National Broadband Plan (NBP), which made significant recommendations for improving access to broadband communications across America and for enhancing the role of broadband in public safety and emergency response.

In particular, the NBP proposed a comprehensive strategy for creating a nationwide interoperable public safety broadband wireless network ("public safety broadband network") for first responders and other public safety personnel. This strategy includes:

* This is an edited, reformatted and augmented version of a Federal Communications Commission publication, dated May, 2010.

- Creating an administrative system that ensures access to sufficient capacity on a day-to-day and emergency basis;
- Ensuring there is a mechanism in place to promote interoperability and operability of the network; and
- Establishing a funding mechanism to ensure the network is deployed throughout the United States and has necessary coverage, resiliency and redundancy.

In this paper, the Omnibus Broadband Initiative (OBI) provides support for the NBP's public funding recommendations for the nationwide interoperable public safety broadband wireless network. This paper also explains how public safety agencies can leverage the deployment of 4G commercial wireless networks to greatly reduce the overall costs of constructing their nationwide broadband network.

INTRODUCTION

The NBP's vision is to create a communications system that allows public safety agencies to take full advantage of cutting-edge broadband technologies. It is therefore essential that public safety agencies have access to commercial technologies, ruggedized for public safety use. This leveraging of commercial technologies will enable public safety agencies to achieve greater communications capabilities, but at much lower costs.

The NBP's vision for the future of public safety broadband communications encompasses several elements:

As shown in Exhibit 1, a multi-pronged approach will provide public safety with greater dependability, capacity and cost savings. First, the hardened network will provide reliable service throughout a wide area. Second, since emergency responders will be able to roam on commercial networks, capacity and resiliency will improve (at a reasonable cost). Third, localized coverage will improve through the use of fixed microcells and distributed antenna systems (DAS)—like those that provide indoor coverage in skyscrapers. Fourth, equipment can be retrieved from caches and used during a disaster when infrastructure is destroyed, insufficient or unavailable, and fire trucks, police cars and ambulances can become mobile picocells.[1]

The NBP requests total public funding to support the construction and ongoing costs of the public safety broadband network. The total present value of the capital expenses and ongoing costs for the network over the next 10 years is approximately $12-16 billion. State and local governments could contribute funds to cover some of these costs, and there may be additional cost-saving methods that reduce this estimate—such as sharing federal infrastructure, working with utilities or use of state and local tower sites.

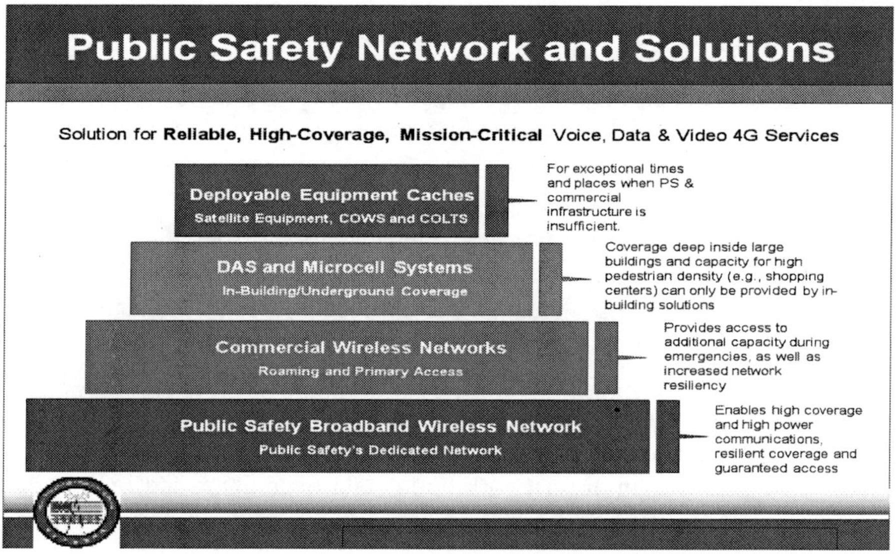

Exhibit 1. The Future of Public Safety Broadband Communications.

The NBP proposes the creation of a public funding program of as much as $6.5 billion capital expenses (capex) in constructing the public safety broadband network. Public funding will be targeted at constructing a public safety overlay network that exploits existing commercial and public safety narrowband infrastructure, as well as: expanding rural coverage; strengthening existing infrastructure; and developing an inventory of deployable equipment. To ensure interoperability, the funding agency should condition all funding awards on compliance with Emergency Response Interoperability Center's (ERIC) requirements.

The public funding program is designed to achieve nationwide interoperability while preserving a great deal of local flexibility. Although

ERIC will set common standards and practices for the nationwide network, public safety agencies at the regional or local level may issue Requests for Proposal (RFPs) and then voluntarily enter into contract with the commercial partners of their choice. This approach will empower each region or locality to satisfy its unique communications needs while promoting vigorous competition among commercial operators and systems integrators for public safety customers.

The NBP also suggests a public funding method, such as imposing a minimal public safety fee on all broadband users, to fund the network's ongoing costs, which include operating expenses (opex) and appropriate network improvement costs. The public funding agency should be charged with disbursing these funds, and any use of such funds must contribute to the operation or evolution of the network and comply with ERIC requirements.

The cost model the NBP used to calculate capital expenses and ongoing costs for the network and to inform its recommendation for the public funding program was validated through multiple approaches.[2] First, a detailed radio frequency (RF) model was constructed, and its RF assumptions were validated through a technical analysis that used data acquired from several major commercial service providers, their competitors and vendors. Costs were based on appropriate comparables, including tariff rates, actual proposals from service providers for similar network builds and operations, and information obtained directly from service providers, equipment vendors, and integrators. Detailed cost scenarios were also developed—and compared with cost scenarios provided by service providers and equipment vendors—to further validate costs.[3]

Assumptions

The NBP's proposal for a public safety public funding program is designed pragmatically to ensure achievement of high-quality public safety broadband wireless service. The planned network focuses on data and video service initially. Over time, it will support wireless voice services used routinely by first responders, and eventually the specialized voice services provided to first responders via the land mobile radio (LMR) service today. The model assumes data and video services via IP transport in the early years, evolving to the target of interoperable mission-critical voice, data and video IP networks and applications in the long term, supported by necessary innovations for mission-critical service.

An incentive-based partnership model is assumed for the estimates, (except under Section E), under which public safety network operators will partner with commercial operators or systems integrators to construct and operate the network using the 10 megahertz of 700 MHz public safety broadband spectrum. Under this model, the vast majority of sites will be built by a commercial partner, either a wireless operator, equipment vendor or a system integrator. The model assumes a 700 MHz Long-Term Evolution (LTE) network. Costs include installing and operating the dedicated 700 MHz Radio Access Network (RAN) and sharing back-haul and IP core transport systems, including ancillary and support systems and services. The IP network architecture enables public safety agencies to have their own dedicated servers for applications and services requiring high levels of security and privacy. The projected costs are not discounted for competitive bidding dynamics, such as strategic value to RFP respondents.[4]

The model assumes that the 10 megahertz of 700 MHz public safety broadband spectrum will be "lit" using LTE technology by exploiting commercial infrastructure, which would result in significant cost and operating efficiencies. LTE commercial rollout is planned with availability to 95% of the United States population by 2015.[5] The public safety capability will be added to this network with targeted site upgrades. The network will be built to support standard commercial devices that operate at low power levels of 23 dBm (decibels of the measured power to 1 milliwatt). In-building penetration loss assumptions are assumed for the non-rural population areas. Public safety will then be able to achieve better coverage and performance than commercial systems by using higher-gain devices with specialized antennas. For highly rural areas, the cost model assumes deployment of a network to support vehicular coverage with externally mounted antennas (EMA) to achieve 99% population coverage.[6] Cell sites in highly rural areas are accounted for as a blend of sites built on existing structures and new sites. Hardening for all sites is also accounted for in the model,[7] and the model further assumes that deployable caches of equipment will be available for emergency use.[8]

Ongoing costs were also calculated on the basis of an incentive-based partnership model. This model assumes that backhaul, core network, managed IP services and ancillary services will be paid through an operating expense charged through a managed service fee. This managed service fee is based on the existing air card managed service fee structure— with the radio access network (RAN) share of the service eliminated, since public safety partners will be using their own spectrum for their primary service.[9]

There are several factors that result in lower capacity requirements for the core network. These include roaming on commercial wireless networks, priority wireless service on commercial broadband 700 MHz networks, deployables (e.g., next generation cells on wheels (COWS) and cells on light trucks (COLTS)) and in-building supplementation, which provide resiliency for capacity surges, increased coverage and increased redundancy.

Capital Expenses (Capex)

As much as $6.5 billion in capital funding will be required over 10-year period to provide advanced public safety broadband network capabilities to agencies that collectively serve 99% of all Americans.

The 10-year estimate of $6.5 billion in capex was developed based on the following assumptions (see Exhibit 2):

- $4.0 billion to equip 41,600 commercial towers with dedicated public safety broadband spectrum RAN capabilities;
- $1.5 billion to harden the commercial towers (improving reliability, particularly when commercial power is lost);
- $0.8 billion to equip 3,200 rural towers with public safety broadband spectrum RAN capabilities by upgrading towers (75%) and installing and equipping new towers (25%) and hardening those towers; and
- $0.2 billion to provide for a fleet of public safety deployables (a mix of next generation COWS, COLTS, etc.), vehicular area network systems and non-recurring engineering costs for handset development.[10]
- Based on this model, a reasonable year-by-year projection of capital expenses is depicted in Exhibit 3.[11]

Ongoing Costs

As previously noted, public funding, such as broadband user fees, will fund the ongoing costs of the network and the network evolution.[12] Following a ramp-up coinciding with the network's expansion, the cost of funding operating costs will reach approximately $1.3 billion per year by the 10th year of construction.

Exhibit 2. Capex Chart

Item	Cost	Notes
41,600 Commercially Deployed Non-rural Sites	$4.0 B	Excludes hardening costs Ethernet over fiber backhaul connectivity to commercial carrier's backhaul Assumes PS RAN (lit) added to 100% of sites (conservative)
Hardening of Existing Commercial Sites	$1.5 B	Assumes 100% of sites need hardening (conservative)
3,200 Rural Sites (includes hardening)	$0.8 B	Assumes EMA, blend of 25% new and 75% upgraded sites
Deployable Equipment and Development	$0.2 B	COLTS, COWS, vehicular area Distributed systems , NRE for handset development, etc.
TOTAL CAPEX	$6.5 B	

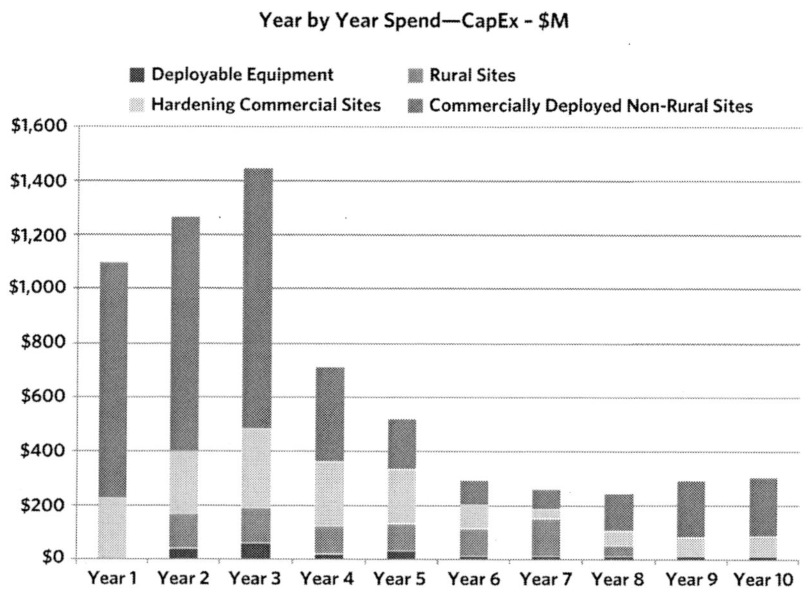

Exhibit 3. Annual Capex Projection.

The $1.3 billion figure was arrived at on the basis of the following assumptions (see Exhibit 4):

- $0.9 billion for IP Managed Services and Transport including backhaul and core from commercial operators exclusive of opex for the public safety RAN;

- $0.2 billion for Managed Services for the dedicated public safety RAN;
- $0.2 billion for additional ongoing costs for rural areas (microwave backhaul, additional site lease cost, etc.); and
- $0.025 billion for operations support for deployable equipment.

In addition, the Plan suggests that this fund be reviewed on a regular basis. Part of this review should also consider whether additional funding is required for network upgrades.

Cost of Separate Public Safety Network

In this section, we compare costs incurred with an incentive-based partnership as described in Section B and costs incurred when an entirely separate dedicated system (stand-alone network) is built for public safety. While the cost estimates for the incentive-based partnership are based on extensive analysis, the costs of the stand-alone network described here are less detailed, in part because of the potential range of ongoing costs. The comparative analysis results in a $6.3 billion capital cost for the network under the incentive-based partnership approach as compared to a $15.7 billion capital cost for a stand-alone public safety network. The cost comparison for these two approaches for both capital and operating costs is even more extreme.

Exhibit 4. Ongoing Network Costs Chart

Item	Cost	Notes
Annual OA&M Including Transport Managed Services Fee	$0.9 B	For 3 million Public Safety Subscribers at $25 per month
Annual RAN Managed Services Fee	$0.2 B	44,800 Sites at $1500 per year for site equipment, OA&M, and $2400 for additional lease cost (this achieves a 99% population coverage)
Additional costs in rural areas (microwave backhaul, additional site lease costs, deployable OpEx)	$0.2 B	Microwave antenna, power and maintenance lease; miscellaneous ongoing costs
TOTAL ONGOING COSTS	$1.3 B	

The technical requirements and capabilities under both approaches are identical and consistent with the assumptions of this paper. Thus, the total number of cell sites remains 44,800.[13] In an incentive-based partnership, we must consider the marginal cost of adding a new radio access network for public safety to an existing tower or site, which already has backhaul to a functioning core network. While it may be necessary to harden the tower or site, many functions can be leveraged. In contrast, for a stand-alone network, we must estimate the full cost for public safety capabilities rather than just incremental costs. The differences emerge in the cost per cell site in both capex and opex; the costs in zoning and site acquisition, because of the need for many more new cell sites beyond the base required for public safety LMR networks; the costs of backhaul from the cell sites; and the costs for a core network.

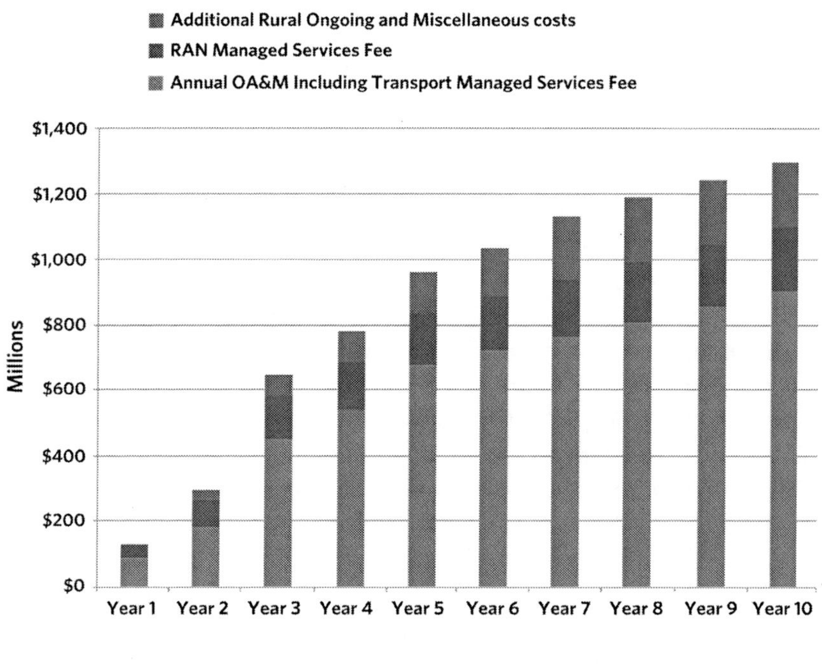

Exhibit 5. Ongoing Costs -Ramp Up (EXAMPLE).

In this analysis, we considered the complexity and scope of constructing a nationwide stand-alone public safety network, in which 80% of the 44,800 sites would be new builds. To avoid comprehensive due diligence requirements and to reduce development costs and time to market, wireless carriers and public safety agencies generally prefer to locate on existing structures rather than build new towers. However, public safety sites must be suitable from a zoning perspective. In many jurisdictions, especially in suburban and rural areas, towers are allowed only on commercially or industrially zoned parcels. Some areas allow towers at agriculturally zoned locations, but most do not allow towers on residentially zoned land, forest land or restricted areas. In addition, sites must not have conditions—such as rocky soil conditions, wetlands, impenetrable trees, possible hazardous waste on properties, high voltage power lines and significant distance to the cell tower site from the main road where utilities are located—that would make constructing a tower extremely expensive. Landowners must also be willing to lease sites at acceptable rates.

Therefore, we assumed that, in urban areas, there are many different antenna sites, such as roof top locations, that public safety agencies can leverage. In suburban and rural America, however, new site acquisition, zoning and construction will in general be substantively higher.

Our analysis indicates that a stand-alone public safety network would be substantially more expensive than a network constructed under the incentive-based partnership approach. Conservatively, the stand-alone network would require at least 2.5 times more capex, excluding deployable equipment, and proportionally even more in ongoing costs.[14] The total present value of the capital expenses and ongoing costs for the standalone network over the next 10 years is approximately $34.4 billion, taking into consideration that capex is $15.7 billion and ongoing costs are 1.5 times the total capex amount.[15] This analysis is consistent with both the Verizon study for the Southern Governors Association, which posited $19 billion for initial capex and total costs of $61 billion over 10 years for capex and ongoing operations,[16] and publicly available information about the costs of New York City NYCWiN broadband network.[17] These results are not surprising given that the incentive-based partnership approach leverages the commercial assets of cellular firms that have large economies of scale by serving 40-100 million customers. By contrast a separate public safety network would not be able to leverage the same assets nor have the same economies of scale, since it would effectively serve only a few million first responders while providing similar nationwide coverage. Further, a separate public safety network does not have similar economies of scope, such as sharing an IP core network with other uses.

This lack of scope is compounded if the public safety entity is operating on an LTE network that utilizes spectrum in a band class assigned exclusively for the public safety community. This would be the case if the D block was reallocated to public safety. In that situation, there would be no commercial service provider in LTE Band Class 14 in the 700 MHz band. While technically such a system could be deployed and supported, the costs of the network equipment, most notably the devices, would increase substantially. Without the ability to leverage the economies of scale of a commercial deployment in a band class, there is significantly less market incentive to develop network equipment and devices capable of operating in that band. Therefore, public safety would have to pay significant premiums for equipment and devices under such a scenario.

Exhibit 6 compares the costs of these two approaches. Overall, the partnership reduces capex and opex by at least 60%.

Comparison Cost of 44,800 Sites		
	Incentive-Based Partnership	Stand-Alone Public Safety Network
Urban Upgraded Site	$95,000	$163,752
Urban New Site	N/A	$223,752
SubUrban Upgraded Site	$95,000	$213,752
SubUrban New Site	N/A	$288,752
Rural Upgraded Site	$216,000	$247,232
SubUrban New Site	$363,000	$394,632
Total CapEx for Sites including Hardening	$6.3 B	$12.6 B
Backhaul - Installation to Core Fiber Ring, Non-Rural Sites	$0	$2.1 B
IP Core Equipment, Network Operations Centers	$0	$1 B
TOTAL CAPEX	$6.3 B	$15.7 B

Exhibit 6. Incentive-Based Partnership vs. Stand-Alone Public Safety Network Capital Expenses.

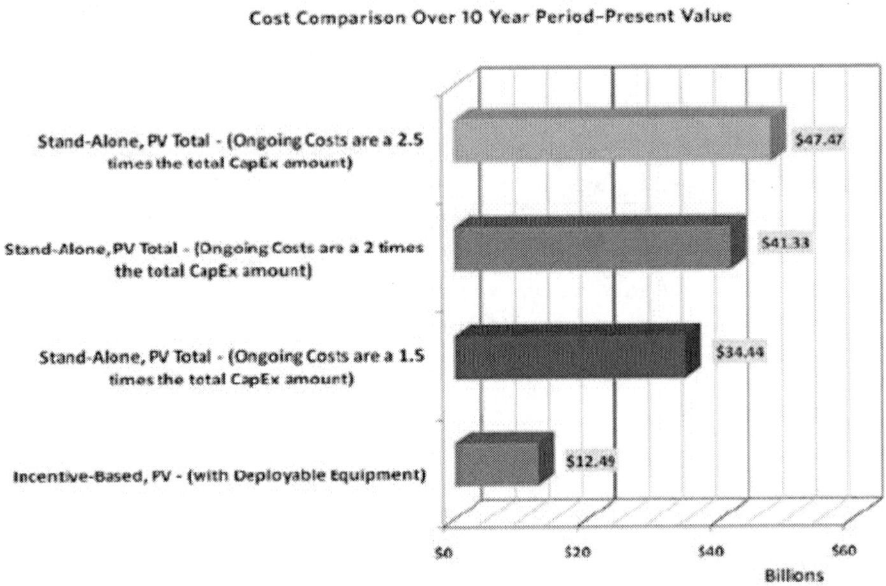

Exhibit 7. Present Value Cost Comparison.

Exhibit 7 provides a cost comparison over a 10-year period for capital and on-going expenses. It shows that the total present value of the capital expenses and ongoing costs for the standalone network over the next 10 years would be approximately $41.3 billion or $47.5 billion—with capex at $15.7 billion and ongoing costs at either two or 2.5 times the total capex amount.[18]

APPENDIX A: DEPLOYABLE EQUIPMENT

The public funding program includes funding for two distinct use cases of public safety deployables:

1. Rapidly deployable full cellular systems that can be deployed for public safety use when either:
   a) A natural disaster or other emergency has occurred in a remote area where there is no public safety 700 MHz cellular system (*e.g.*, a train crash with chemical spills in a remote area or a forest fire in a wilderness area); or

b) The working public safety cellular system for a cell site or larger area has been destroyed or is temporarily inadequate. The systems deployed in such circumstances are sometimes referred to as Cells on Wheels (COWs) and Cells on Light Trucks (COLTs). LTE enables a new generation of this equipment that will be much lighter than current equipment.[19]

2. Vehicles equipped with technology that enables the first responder occupants of the vehicle to use the vehicle communications systems as a relay connecting their handheld to a remote base station. When the officer leaves the vehicle to go into a building or to the physical site of accident (e.g., to investigate a car rolled over an embankment or to pursue a suspect on foot), the handheld device communicates back to the vehicle, which in turn relays the communications back to the closest cellular tower—which may be reachable only from a high-gain vehicle. In effect, the vehicle becomes a vehicular area network (VAN).

The deployable caches included in the public funding will serve all major metropolitan areas and will include sufficient fleets for each state to ensure adequate deployment to reach any emergency within a small number of hours.

This part of the public funding program also includes money for Non-Recurring Engineering costs for the specialized chipset and software development to enable the development of public safety LTE devices in the market that take advantage of commercial capabilities and also ensure the development of any specialized needs for public safety devices. For example, public safety devices must operate in Band Class 14 and be able to roam into other LTE 700 MHz band classes.

APPENDIX B: NETWORK COSTMODEL ASSUMPTIONS

- Network Build Model:
- A pragmatic approach that achieves high quality wireless broadband service using spectrum dedicated for public safety—the 5+5 MHz public safety broadband spectrum—to provide public safety with a dedicated Radio Access Network (RAN).

- Assumes that public safety agencies on an area-by-area basis will collectively issue a Request for Proposal (RFP) for that area for the building out of the public safety broadband network.
- Potential partners: The respondents to the RFP may include any of the following:
- A commercial wireless operator with an existing network, particularly a Long Term Evolution (LTE) network in the geographic area with 700 MHz spectrum (other than the D Block) that adds equipment to "light-up" the public safety broadband spectrum;
- A commercial wireless operator who is a D Block auction winner and is simultaneously building out the LTE Band 14 profile that includes both D Block and public safety spectrum; or
- A systems integrator who is participating by itself or building out as part of an Land Mobile Radio (LMR) or other build for public safety that builds a broadband wireless network only for the public safety broadband spectrum.
- The lowest-cost build would be the synchronous build with the D Block, while the highest cost build would be a stand-alone build by a systems integrator.
- Funding is based on an asynchronous build where existing operators' infrastructure would be expanded to include the "lighting" of the public safety 700 MHz broadband spectrum to give public safety a dedicated RAN. > Assumes LTE commercial rollout availability to 95% of the population will be achieved by market forces by 2015. > For the 95% that are likely to be served by LTE-based operator plans, this would be an asynchronous expansion by an operator who has built out an LTE network. > For highly rural America, where there is not market commitment for an LTE network, build out was modeled to use 2G infrastructure plus new towers where necessary.
- Subscriber device model:
- Commercial power levels (23 dBm) for handheld devices, except in highly-rural areas. Public safety agencies can choose to equip their officers with slightly larger handheld devices with small external antennas and larger batteries, thus gaining 2 to 3 decibels (dBs) of additional power. These devices will provide public safety officers with superior coverage and high speed near cell edges.
- In highly rural areas the subscriber device supported by the network is a vehicular device using an externally mounted antenna (EMA).

Commercial handheld devices will also work in these areas for much of the area within a cell site, but at reduced speeds as one gets closer to the cell edge.

- The model contains no device funding for handheld or the vehicular device with the EMA, as that was assumed to be the responsibility of each individual agency.
- The subscriber devices should be substantially lower in costs than they are today for public safety because of the ability to leverage the commercial device ecosystems. In the operating system, the baseband chipset and the RF chipset are the components of the device that require high volumes to drive costs down. These components will also be used in commercial deployments and thus will be in high volume.
- Network services:
- Data and video services via IP Transport in early years offering a more reliable, high performance, and more cost-effective version of the commercial wireless aircard services that some public safety officers purchase today.
- Commercial voice via VoIP over LTE in the medium term as that becomes available on LTE networks.
- Interoperable, mission-critical voice, data and video IP networks and applications as the long-term target.
- Link budget assumptions:
- In-building penetration loss assumptions are the same as commercial LTE except for highly-rural, which is modeled for vehicular EMA coverage. As noted above, public safety officers can achieve performance superior to commercial performance with handhelds with small external antennas.
- LTE Commercial Speeds with 95% area coverage (256 Kbps uplink typically) can be achieved on top of an LTE commercial service cell site infrastructure with minimal site supplementation.
- Vehicular coverage for highly rural areas to achieve 99% population coverage.
- Grant funding:
- Public funding for paying for the RFPs is based on a commercial winning bidder installing and operating a dedicated public safety broadband 700 MHz RAN that shares backhaul, IP Core transport systems, including ancillary and support systems and services. Public safety agencies may choose to operate dedicated servers for specific applications and services that contain sensitive information.

- Funding is based on the full costs of dedicated RAN build. There is no discount of the prices included for competitive bidding dynamics, such as strategic value to RFP respondents, although such discounts are likely.
- Operating expense assumptions:
- Backhaul, core network and managed IP services and ancillary services provided via wireless operator or systems integrator and paid through opex charged for a managed services fee.
- Managed service fee based on 2010 aircard managed service fee structure with RAN share of service eliminated.
- Annual opex fee incurred for management and maintenance of public safety broadband 700 MHz RAN.
- Capital expense assumptions:
- Cell sites in rural America are treated as a blended build of new sites on existing structures and new sites.
- $95,000 blended average per site capex for adding public safety broadband to commercial LTE cell site.
- $35,000 hardening per site for commercial LTE sites. > $216,000 average per site capex for adding public safety broadband to existing sites in most rural areas, including $75,000 per site for hardening.
- $363,000 average per site capex for public safety broadband new sites in the most rural areas, including $75,000 per site for hardening.
- Priority wireless service on commercial networks, deployables and in-building supplementation provides for capacity surges, more extensive coverage and more resiliency, thus lowering site requirements on the core network.
- The model will be refined based on real-life experience in future public funding years.

APPENDIX C: UNDERLYING EQUIPMENT AND COST FOR CAPITAL EXPENSE ASSUMPTIONS

Equipment and Costs for Blended Average Per Site Capex for Adding Public Safety Broadband to Commercial LTE Cell Sites

Non-Rural Site Configuration A and B, for Asynchronous Build Two different types of configurations (A and B) are used for the underlying equipment for adding public safety broadband to commercial LTE cell sites. In addition, structure heights, or distances from the eNodeB to Antennas for the site locations, were evaluated for cost at 75 feet and 150 feet. The main differences between configuration A and B are that configuration A uses rigid coax and configuration B uses fiber and remote radio heads (RRH). Configuration A uses rigid coax from the eNodeB at the base of the structure/tower up to the top of the tower or structure/tower where the antennas are located. Configuration B uses fiber from the eNodeB at the base of the structure/tower up to the top of the tower or structure/tower where the antennas and RRH are located.

Equipment and Costs Per Site Capex for Adding Public Safety Broadband to Existing Sites in Highly Rural Areas, Including Hardening

Rural Site Configuration A and B, for Asynchronous Build Two different types of configurations are used for the underlying equipment for adding public safety broadband to highly rural areas. In addition, structure heights, or distances from the eNodeB to Antennas for the site locations, were evaluated at 225 feet. Microwave equipment and hardening are also included in the underlying cost analysis.

Equipment and Costs Per Site Capex for Public Safety Broadband New Sites in Highly Rural Areas, Including Hardening

Two different types of configurations (A and B) are used for the underlying equipment for new sites in highly rural areas. In addition, structure heights, or distances from the eNodeB to antennas for the site locations, were evaluated at 225 feet. Microwave equipment and hardening was also included in the underlying cost analysis.

New sites in highly rural areas also included Site Acquisition and Construction of up to a 225 foot structure/tower.

Non-Rural Site Configuration A, for Asynchronous Build

Non-Rural Site configuration for Asynchronous Build where existing operators infrastructure would be expanded to include the "lighting" of the public safety 700 MHz broadband spectrum to give public safety a dedicated RAN.

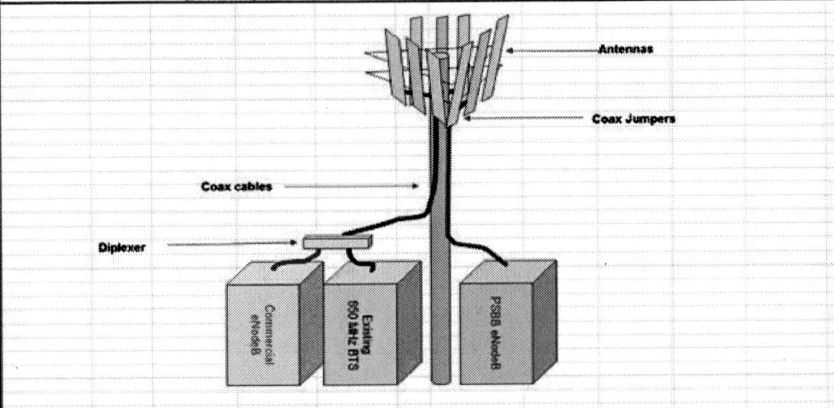

Configuration A

75 Foot site

			Per Coax run	Per sector (2 Coax runs)
1/2" Low-Loss Foam Coax	$0.65	per foot	$58.50	$117.00
CONNECTOR 1/2" ONE PIECE RAPID FIT DM	$16.00	each	$32	$64.00
Coax Jumper, 6 foot, SCF12-50J	$54.00	each	$108	$216.00
MIMO Antenna and mounts	$2,250.00	each	$2,250	$2,250
1/2 inch GROUND KIT (CLAMP) FOR COAX	$12.00	Each coax run	$12	$24.00
Coax Clamps	$25.00	2 pkg/coax run	$50	$100.00
CUSHION HANGER ASSEMBLY 1/2 COAX (PACK OF 5)	$24.00	3 pkg/coax run	$72	$144.00
Tower Mounted Amplifier (TMA)	$1,550.00	each		
		Total above	$2,582.50	$2,915.00
			Configuration A	
		Tower Equipment	3 Sector Cost	$8,745.00
			eNodeB Termination Fiber Ethernet	$3,000
			eNodeB	$45,000
			Total Equipment	$56,745
			40% Eng.& Installation	**$22,698.00**
			Grand Total	$79,443.00

150 Foot site

			Per Coax run	Per sector (2 Coax runs)
7/8" Ultraflexible Foam Coax	$1.18	per foot	$194.70	$389.40
CONNECTOR 7/8" ONE PIECE RAPID FIT DM	$17.00	each	$34	$68.00
Coax Jumper, 6 foot, SCF12-50J	$58.00	each	$116	$232.00
MIMO Antenna and mounts	$2,250.00	each	$2,250	$2,250
7/8 inch GROUND KIT (CLAMP) FOR COAX	$12.75	Each coax run	$13	$25.50
Coax Clamps	$27.00	2 Each coax run	$54	$108.00
CUSHION HANGER ASSEMBLY 7/8 COAX (PACK OF 5)	$25.50	3 each coax run	$77	$153.00
Tower Mounted Amplifier (TMA)	$1,550.00	each	$1,550	$3,100.00
		Total above	$4,287.95	$6,325.90
			Configuration A	
		Tower Equipment	3 Sector Cost	$18,977.70
			eNodeB Termination Fiber Ethernet	$3,000
			eNodeB	$45,000
			Total Equipment	$66,978
			40% Eng.& Installation	**$26,791.08**
			Grand Total	$93,768.78

Non-Rural Site Configuration B, for Asynchronous Build

Non-Rural Site configuration for Asynchronous Build where existing operators infrastructure would be expanded to include the "lighting" of the public safety 700 MHz broadband spectrum to give public safety a dedicated RAN.

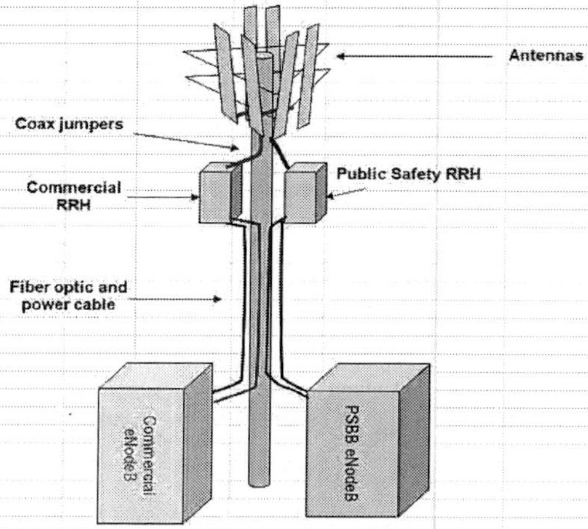

Configuration B

75 Foot site						150 Foot site	
				Per Fiber run	Per sector (2 Fiber runs)	Per Fiber run	Per sector (2 Fiber runs)
Coax Jumper, 6 foot, SCF12-50J	$54.00	each		$108	$216.00	$116	$232.00
MIMO Antenna and mounts	$2,250.00	each		$2,250.00	$2,250	$2,250.00	$2,250
Remote Radio Head (RRH)	$4,875.00	each		$4,875	$4,875	$4,875	$4,875
Fiber Cable	$0.30	per foot		$27.00	$54	$49.50	$99
Power Cable	$0.05	per foot		$4.50	$9	$8.25	$17
Fiber Clamps	$11.00	2 pkg/run		$11.00	$22	$11.00	$22
Fiber connectors	$3.50	each		$7	$14	$7	$14

	75 Foot site		150 Foot site	
Total above	$7,282.50	$7,440.00	$7,316.75	$7,508.50
	Configuration B		Configuration B	
Tower Equipment	3 Sector Cost	$22,320.00	3 Sector Cost	$22,525.50
	eNodeB Termination Fiber Ethernet	$3,000	eNodeB Termination Fiber Ethernet	$3,000
	eNodeB	$45,000	eNodeB	$45,000
	Total Equipment	$70,320	Total Equipment	$70,526
	40% Eng.& Installation	$28,128.00	40% Eng.& Installation	$28,210.20
	Grand Total	$98,448.00	Grand Total	$98,735.70

Rural Site Configuration A, for Asynchronous Build

Rural Site configuration for Asynchronous Build where existing operators infrastructure would be expanded to include the "lighting" of the public safety 700 MHz broadband spectrum to give public safety a dedicated RAN.

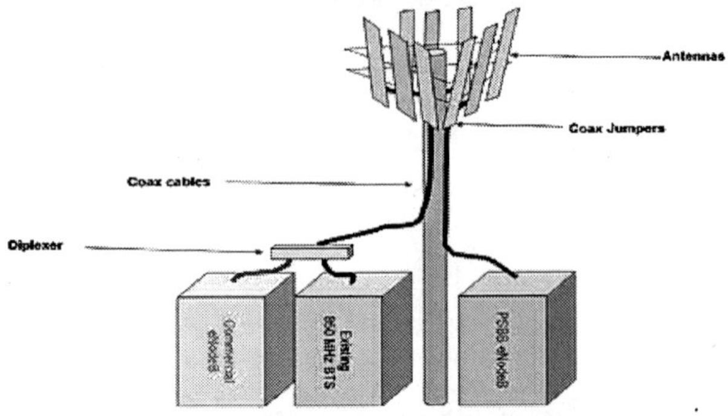

Configuration A

225 Foot site				
			Per Coax run	Per sector (2 Coax runs)
1-1/4" Premium Attenuation Coax	$2.95	per foot	$708.00	$1,436.00
CONNECTOR 1-1/4" ONE PIECE RAPID FIT DM	$17.00	each	$122	$243.00
Coax Jumper, 6 foot, SCF12-50J	$58.00	each	$116	$232.00
MIMO Antenna and mounts	$2,250.00	each	$2,250	$2,250
1-1/4" GROUND KIT (CLAMP) FOR COAX	$12.75	Each coax run	$19	$17.00
Coax Clamps	$27.00	2 Each coax run	$140	$280.00
CUSHION HANGER ASSEMBLY 1-1/4" COAX (PACK OF 5)	$25.50	3 each coax run	$201	$402.00
Tower Mounted Amplifier (TMA)	$1,550.00	each	$1,550	$3,100.00
		Total above	$5,105.00	$7,960.00
			Configuration A	
		Tower Equipment	3 Sector Cost	$23,880.00
			eNodeB termination Fiber Ethernet	$3,000
			eNodeB	$45,000
			Total Equipment	$71,880
			40% Eng.& Installation	$28,752.00
			Grand Total	$100,632.00

Total Rural Upgrade Site	Cost
Hardening	$70,000
Microwave equipment	$45,000
3 sector Equipment and Installation	$100,632.00
Total:	$215,632

Total New-Build Rural Site	Cost
Hardening	$70,000
Microwave equipment	$45,000
Site Acquisition, Zoning	$57,400
Construction	$90,000
3 sector Equipment and Installation	$100,632.00
Total:	$363,032

Rural Site Configuration B, for Asynchronous Build

Rural Site configuration for Asynchronous Build where existing operators infrastructure would be expanded to include the "lighting" of the public safety 700 MHz broadband spectrum to give public safety a dedicated RAN.

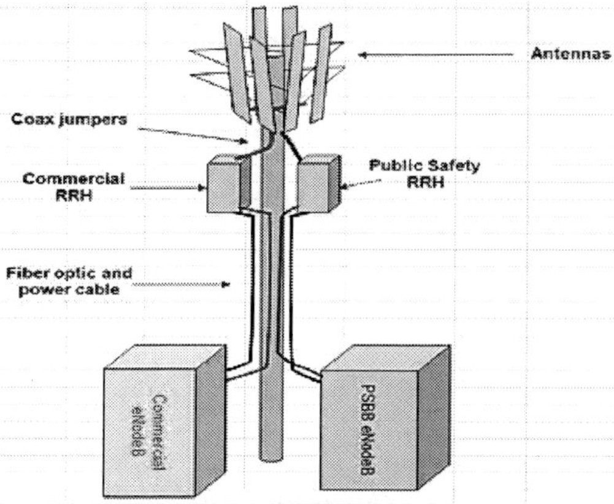

Configuration B

225 Foot site					
				Per Fiber run	Per sector (2 Fiber runs)
Coax Jumper, 6 foot, SCF12-50J	$54.00	each		$116	$232.00
MIMO Antenna and mounts	$2,250.00	each		$2,250.00	$2,250
Remote Radio Head (RRH)	$4,875.00	each		$4,875	$4,875
Fiber Cable	$0.30	per foot		$72.00	$144
Power Cable	$0.05	per foot		$12.00	$24
Fiber Clamps	$11.00	2 pkg/run		$11.00	$22
Fiber connectors	$3.50	each		$7	$14
			Total above	$7,343.00	$7,561.00

	Configuration B	
Tower Equipment	3 Sector Cost	$22,683.00
	eNodeB Termination Fiber Ethernet	$3,000
	eNodeB	$45,000
	Total Equipment	$70,683
	40% Eng.& Installation	$28,273.20
	Grand Total	$98,956.20

Total Rural Upgrade Site	Cost
Hardening	$70,000
Microwave equipment	$45,000
Equipment and	$98,956.20
Total:	$213,956

Total New-Build Rural Site	Cost
Hardening	$70,000
Microwave equipment	$45,000
Site Acquisition, Zoning	$57,400
Construction	$90,000
3 sector Equipment and Installation	$98,956.20
Total:	$361,356

HARDENING COST

Rural

CAPEX	Item	Cost
	Additional Batteries and cabinet (8 to 12 hours minimum additional battery back-up)	$4,800
	20 KW Diesel Generator with 200 AMP Transfer Swich, CLIFFORD	$11,425
	Electrical wires to eNodeB and tower	$325
	1,000 gallon fuel tank	$1,750
	Fuel	$2,000
	Fuel Lines	$250
	Improve antenna survivability designed for a wind loading - 100 to 140 mph (The Electronics Industry Association Structural Standards for Steel Antenna Tower and Antenna Supporting Structures - EIA/TIA-222)	$5,500
	Improve Structural Standards for Steel Antenna Towers (The Electronics Industry Association Structural Standards for Steel Antenna Tower and Antenna Supporting Structures - EIA/TIA-222)	$25,750
	Structural Analysis per base station	$2,500
	Installation (40%)	$20,720
	Total:	**$75,020**

Non-Rural

CAPEX	Item	Cost
	Additional Batteries and cabinet (8 hours minimum additional battery back-up)	$4,250
	Electrical wires to eNodeB and tower	$200
	Improve antenna survivability designed for a wind loading - 100 to 140 mph (The Electronics Industry Association Structural Standards for Steel Antenna Tower and Antenna Supporting Structures - EIA/TIA-222)	$3,500
	Improve loading and/or Structural Standards for structure or Towers (The Electronics Industry Association Structural Standards for Steel Antenna Tower and Antenna Supporting Structures - EIA/TIA-222)	$10,050
	Miscellaneous	$5,250
	Structural Analysis per base station	$2,500
	Installation (40%)	$9,300
	Total:	**$35,050**

	Optional Equipment not included	
	20 KW LP Generator with 200 AMP Transfer Swich, CLIFFORD	$7,900
	Portable, 8000 WATT GENERATOR, Transfer Switsh, GENERAC	$4,500

MICROWAVE SYSTEM (cost per 100 Mbps Ethernet terminal)

Item	Cost
Microwave installation (6 and 11 GHz)	$14,500
1/2" Low-Loss Foam Coax ($0.65 per foot)	$156
CONNECTOR 1/2" ONE PIECE RAPID FIT DM	$32
6 foot, SCF12-50J	$108
Coax Clamps	$75
CUSHION HANGER ASSEMBLY 1/2 COAX (PACK OF 5)	$100
Microwave WG, Hydrator and Accessories (6 and 11 GHz)	$4,500
FCC Application and Coordination	$1,000
Microwave Antenna (6 or 11 GHz)	$2,500
Microwave Terminal HSB (6 and 11 GHz)	$17,300
Path Survey	$1,750
Zoning	$3,000
Total:	$45,021

Hardening

Hardening includes additional batteries and battery cabinet, structural analysis and improving the cell-site structure and antenna survivability designed for a wind loading, according to the Electronics Industry Association Structural Standards for Steel Antenna Tower and Antenna Supporting Structures (EIA/TIA-222).

For rural sites, hardening also includes adding generators and associated equipment.[20]

Microwave

Microwave equipment includes all equipment, path survey and installation for the microwave system. In addition, FCC applications, coordination and zoning are included in the cost structure.

REFERENCE

Tessco, http://www.tessco.com/ (last visited March, 6 2010).

Meridian Microwave, http://www.meridianmicrowave. com/#0%20-%206% 20miles (last visited March, 6 2010).

Radio Frequency Systems, http://www.lwsonline.de/oxid/ (last visited March, 6 2010).

Huawei, http://www.huawei.com/site_products.do (last visited March, 6 2010).

Powerwave, http://www.powerwave.com/datasheets.asp (last visited March, 6 2010).

Clifford Power Systems, Inc., http://www.cliffordpower.com/ (last visited March, 6 2010).

Fred A. Nudd Corporation, http://www.nuddtowers.com/ (last visited March, 6 2010).

Telecommunications Industry Association, http://www.tiaonline.org/ (last visited March, 6 2010).

Commscope, "Wind Loading on Basic Antennas," http://www.commscope. com/andrew/eng/support_document/ tech_info/antennas/__ics Files/afield file/2009/07/13/wind_ loading_on_base_stationantennas_white_Paper.pdf

Valere Power Systems, http://www.eltekvalere.com/wip4/ (last visited March, 6 2010).

http://www.generac.com/Full_Product_Line/ , (last visited March, 6 2010).

APPENDIX D: CAPEX FOR PUBLIC SAFETY 700 MHZ BUILDS—STAND-ALONE

CAPEX FOR PUBLIC SAFETY 700 MHZ BUILDS - STAND-ALONE	Cost
Urban Cost per cell site for Upgrade	$163,752
Urban Cost per cell site for New Site	$223,752
SubUrban Cost per cell site for Upgrade	$213,752
SubUrban Cost per cell site for New Site	$288,752
Rural Cost per cell site for Upgrade	$247,232
Rural Cost per cell site for New Site	$394,632
% of Sites Upgraded (All Sites)	20%
% of Sites New Build (All Sites)	80%
Base Case Urban Areas, Upgraded Sites	$596,418,409
Base Case Urban Areas, New Sites	$3,259,801,634
Total - Urban	$3,856,220,043
Base Case SubUrban Areas, Upgraded Sites	$998,009,209
Base Case SubUrban Areas, New Sites	$5,392,736,834
Total - SubUrban	$6,390,746,043
Total - Non-Rural	$10,246,966,085
Upgraded Sites w/ EAM in Rural Areas	$155,113,357
New Sites w/ EAM in Rural Areas	$990,368,467
Hardening non-Rural Sites	$1,163,568,000
TOTAL RAN CAPEX FOR STAND-ALONE SCENARIOS	$12,556,015,909
Backhaul - Installation to Core Fiber Ring, Non-Rural Sites	**$2,078,672,676**
IP Core Equipment, Network Operations Centers	**$1,027,939,000**
Grand Total - STAND-ALONE Build for Public Safety	**$15,662,627,585**

End Notes

1 For an extensive discussion of how the public safety broadband network will use deployable equipment, see Appendix A.

2 A detailed discussion of the assumptions underlying the network cost model is provided in Appendix B.

3 Because network designs and assumptions may change over time, it is imperative that the funding agency ensures that there is an annual review of the funding available for each program.

4 Through partnering, RFP respondents will see an effective reduction in capex associated with their own network build out as well as an improvement in reliability through side hardening for public safety.

5 See http://www22.verizon.com/Content/ExecutiveCenter/Richard_Lynch/mobile_world_congress/mobile_ world_congress.htm (last visited Feb. 20, 2010); see also http://www.att.

com /gen/press-room?cdvn=news&news articleid=30493&pid=4800 (last visited Feb. 20, 2010).

[6] The model excludes the costs of EMAs, which are components of subscriber devices used for vehicular coverage. EMAs are standard equipment used in public safety vehicles to improve coverage.

[7] We assumed $35,000 per site for hardening in non-rural areas, and $70,000 per site in highly rural areas.

[8] Portable user equipment or radios with ancillary support equipment stored and available for emergency use.

[9] We have not included any costs that might be incurred for roaming by the public safety operator on a commercial network. Public safety will be able to obtain roaming services at favorable commercial rates.

[10] This assumes a cost range from $100,000 to $400,000 for next generation deployable cell sites as well as a cost of up to $10,000 per vehicle for vehicular area network systems.

[11] Appendix C provides more detail on the cost model used to calculate overall capital expenses for the network. Actual costs for a particular region for a specific RFP will vary on a line-by-line basis.

[12] The proposed funding covers network operations. The funding is not intended to cover the operations of the services and applications running on top of that network nor various administrative functions associated with public safety network operations that agencies may incur. These costs which are part of day-to-day operations today which we have assumed will continue to be borne by the local agencies.

[13] Public safety regions could deploy networks with fewer cell sites, but such networks would provide worse performance, slower speeds, and less total capacity. For the case of the Stand-Alone build we used 20% existing public safety sites and 80% new sites, based on the number of LMR sites that typically serve a region compared with the number of cellular sites.

[14] Based on Sprint Nextel and Verizon Wireless annual reports for 2009, OpEx is approximately twice CapEx. Based on our analysis, the range of ongoing costs is 1.5 to 2.5 times the total CapEx amount, with two times the total CapEx amount as the norm. For some Stand-Alone networks, Ongoing Costs could be higher. For these reasons, we have estimated costs based on a range.

[15] Ongoing Costs equal to 1.5 times the total CapEx amount is the lower bound. Appendix D provides a more detailed cost breakdown of the CapEx for the $15.7 B

[16] See SGA Task Force: Achieving Interoperability for Public Safety Communications (2007); Response of Verizon Communications and Verizon Wireless (Mar. 16, 2007).

[17] See Henry Morgenstern, NYCWiN Interoperable Communications: A Report on the New York City Wireless Network, Counter Terrorist Magazine, Sept./Oct. 2008, available at http:// www.thecounterterroristmag.com/ pdf/Issue3.NYCWiN.Morgenstern.Lo.pdf (last accessed Mar. 26, 2010). See also Department of Information Technology and Telecommunications, Testimony before the City Council Committees on Fire and Criminal Justice Services, Public Safety, and Technology in Government Oversight – Implementation Status of the New York City Wireless Network (Feb. 25, 2008).

[18] See supra 14

[19] Letter from Brian Ponte, Vice President for Business Development, LEMKO Corporation, to Marlene H. Dortch, Secretary, FCC (Mar. 12, 2010).

[20] Many commercial sites today have battery back-up and structural hardening and back-up power systems for primary sites but not for secondary sites. The model assumed hardening

and batteries for all sites with diesel generators as optional. In practice, the funds not needed for sites that are already hardened could be used for diesel generators at other sites. The localization of the RFP approach allows solutions to be tailored to the local needs and environment.

In: Emergency Communications ISBN: 978-1- 62081-140-5
Editors: D. H. Martinez and A. E. Young © 2012 Nova Science Publishers, Inc.

Chapter 4

PUBLIC SAFETY COMMUNICATIONS: ARE THE NEEDS OF OUR FIRST RESPONDERS BEING MET? TESTIMONY OF CHIEF JACK PAROW, MA, EFO, CFO, PRESIDENT AND CHAIRMAN OF THE BOARD, PRESENTED TO THE COMMITTEE ON HOMELAND SECURITY US HOUSE OF REPRESENTATIVES*

International Association of Fire Chiefs

March 30, 2011

Good morning Chairman King, Ranking Minority Member Thompson, and members of the committee. I am Jack Parow, recently retired chief of the Chelmsford (MA) Fire Rescue Department and president and chairman of the board of the International Association of Fire Chiefs (IAFC) on whose behalf I appear. My organization represents the leadership of over 1.2 mllion firefighters and emergency responders. IAFC members are the world's leading experts in firefighting, emergency medical services, terrorism response,

* These remarks were delivered as testimony given on March 30, 2011. Jack Parow is a retired chief of the Chelmsford (MA) Fire Rescue Department and president and chairman of the board of the International Association of Fire Chiefs (IAFC).

hazardous materials spills, natural disasters, search and rescue, and public safety policy. As far back as 1873, the IAFC has provided a forum for members to exchange ideas and find the latest products and services available to first responders.

I testify today in support of H.R. 607 (Broadband for First Responders Act of 2011). Currently, this legislation has eleven cosponsors and is bipartisan. Obviously, this is not part of a political agenda. It is legislation to benefit public safety agencies and the citizens whom they serve and protect. We are grateful for the efforts of this committee to keep this issue foremost in congressional consideration. Public safety is an integral part of this nation's homeland security. Local fire, emergency medical services (EMS) and law enforcement constitute the first response to incidents both small and large--natural, accidental and illicit. This legislation is needed to enable fire, EMS and law enforcement to reach our highest goal – construction of a nationwide public safety wireless interoperable broadband network.

For too long we have had an urgent need to improve public safety communications. Our business is incident management and the fundamentals here are command and control. These elements are not possible without sufficient communications capability. This was acknowledged in both the 9-11 Commission and Katrina reports which summarized the deficiencies of response to those catastrophic events.

Today, public safety communications is served by Land Mobile Radio (LMR) of which there are some 55,000 separate agencies licensed to broadcast. These LMR radio systems are licensed over six or more bands of radio frequencies. Each public safety licensee can operate only in its own geographic area so as not to interfere with a licensee in another jurisdiction on the same frequency. Over the past 50 years, the Federal Communications Commission (FCC) has provided thin slices of spectrum for public safety as each frequency band became available. This arrangement has led to a patch-work of radio systems across the country that are mostly not interoperable which makes it difficult and expensive for agencies in the same or nearby jurisdictions to have interoperability. To solve this problem and promote nationwide interoperability, a national architecture for public safety communications is required to bring public safety communications into the 21st Century. To achieve a nationwide public safety wireless broadband network, key elements need to be in place.

First, the network must have *sufficient capacity*. To achieve a nationwide public safety broadband network – connectivity coast to coast, border to border – 10 MHz of D Block spectrum, currently slated for FCC auction, must be added to our 10 MHz of spectrum already licensed to public safety. That would give us a 20 MHz network. Public safety, under FCC regulation, is required to use Long Term Evolution as its broadband technology. To maximize the potential of this technology, a configuration of 20 MHz is needed. As you can see on the spectrum chart below, the currently licensed public safety spectrum abuts the D Block and is perfect for public safety.

New Upper 700 MHz Band Plan - Adopted by FCC on July 31, 2007

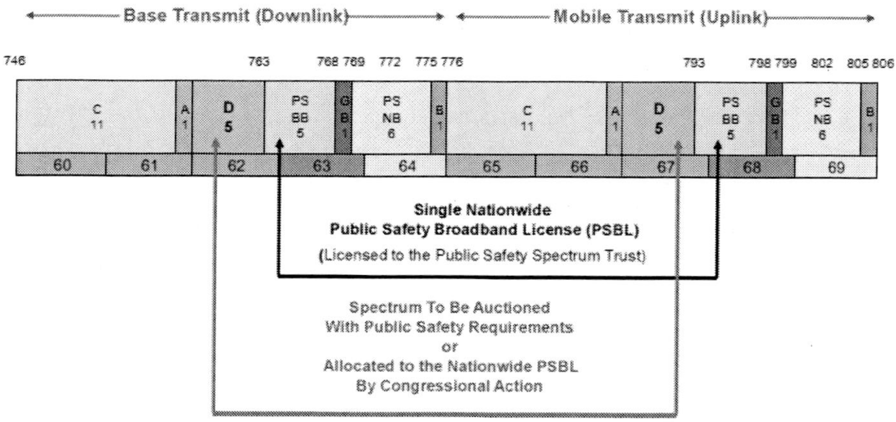

With this configuration only, not with any other, can public safety be assured that it will have the ability to build the network it needs now and into the future. H. R. 607 can do this for us. Here we have a one-time opportunity, one chance to get it right. Secondly, *the network must be under public safety control.* Local control of the network by public safety agencies is critical. A single licensee utilizing a single technology with sufficient spectral capacity will ensure nationwide interoperability and allow us to effectively manage day to day operations, along with major incidents. We cannot have commercial providers deciding what is or is not an emergency or what is the priority. Public safety transmissions must go through without delay. The lives of our firefighters and medics depend on this necessity. A "no service" signal is not accepted in emergency operations.

Public safety expects to enter into public-private partnerships. We will work with state, county and local governmental agencies, federal partners, electric and gas utilities and others who respond to emergencies such as highway and water agencies. However, public safety must have control over the operation of the network in real time. Network control will give public safety certainty that it will have full, immediate, preemptive priority over its spectrum on a when-needed basis. This is a public safety imperative.

Third, *the network must be mission critical at the outset.* In the beginning, this system will handle only data and video. At some future time – years from now – we believe there will be a transition to mission critical voice. We all need to take a long term view which means starting with sufficient spectrum so that we will have the ability to migrate to mission critical voice in the future. This will occur when the technology becomes available, when public safety has confidence in it, and when its cost is affordable. Here are the key elements of "mission critical:"

- The network must be hardened to public safety requirements. This means towers must be able to withstand elements that might disable them. Towers in hurricane-prone areas and tornado alleys must be designed accordingly. Back up electrical power must be available 24/7. Redundancy is necessary.
- The public safety mission critical voice network must have the ability to broadcast and receive "one-to-one" and "one-to-many." It must also have the ability to broadcast and receive without the network infrastructure being operative. This is called the "talk around" mode – also known as simplex. This is a command and control absolute. You know very well that we operate under extremely hazardous conditions. If for any reason the network cannot provide connectivity, then we need the capability to communicate without the network. This means communicating in the simplex mode. Herein lies the very essence of public safety communications.
- The network must have back up capabilities in the event of network loss. We envision satellite capability when a tower is disabled or other crippling malfunction occurs in the network. Satellites also can cover remote areas that do not have terrestrial broadcast facilities. Our mission is geography-oriented whereas commercial carriers are population-oriented.

And, fourth, *funding is important for the construction of a public safety broadband network.* State and local government budgets are challenged. The broadband network needed by public safety cannot be built without federal funding support. H. R. 607 recognizes this fact and offers a solution. According to the proposed legislation, both a Public Safety Interoperable Broadband Network Construction Fund and a Public Safety Interoperable

Broadband Network Maintenance and Operation Fund would be established in the Treasury of the United States. The Secretary of Homeland Security would be required to establish a Construction Grant Program and to administer a Maintenance and Operation Reimbursement Program. All of this is welcome and necessary.

We cannot underestimate how this public safety broadband network will revolutionize the fire and emergency medical services. For example, the network could provide live video to provide instantaneous situational awareness for mass casualty incidents, major hazardous materials spills, and real time situational awareness for incident command as well as elected officials and other decision makers. In the area of emergency medical services, we expect digital imaging, portable EKGs, portable ultrasounds, and field blood work with a direct link to the hospital's emergency department. This would put a virtual physician in the back of the ambulance with the Emergency Medical Technician to expedite the proper life saving treatment enroute to the hospital. These types of applications for fire and EMS are only possible with broadband capability.

One area of the bill which will need attention as the legislation moves forward is Section 207, which mandates migration of public safety entities in the 420 to 512 MHz band to 700/ 800 MHz frequencies. We understand the intent of this provision is to achieve long term interoperability by consolidating band use by public safety. The IAFC supports planned migration to 700/800 MHz frequencies but is concerned with inclusion of mandatory time frames in which to achieve it.

Mr. Chairman, the IAFC wholeheartedly supports H.R. 607. This bill provides public safety with the spectrum and funding to begin the hard work of constructing a nationwide public safety broadband network. H.R. 607 is our vehicle for finally securing this critical resource, and we want to work with you and your colleagues in the House of Representatives to further refine this legislation in order to enact the best possible bill into law. The 10th anniversary of the tragic events of September 11, 2001 is little more than 5 months away. Thus, we urgently need to continue to move forward on a plan to make the vision of a public safety broadband network a reality. Thank you for your personal commitment and leadership on this critical issue. I am available to respond to questions.

INDEX

D

E

F

G

P

R

S